国家出版基金项目
NATIONAL PUBLICATION FOUNDATION

丛书主编：吉日木图

骆驼精品图书出版工程

骆驼组织学彩色图谱

苏布登格日勒　李海军 ◎ 著

中国农业出版社
北　京

内容简介

本书详细描述了骆驼,尤其是双峰驼各系统主要器官组织学结构,根据系统顺序分别介绍低倍与高倍视野下的整体与局部结构。在内容选择上,精选了双峰驼各主要代表性器官的组织切片;在内容表现上,除了在图片上直接标注外,各章前还有一段精炼的有关该系统组织结构的文字概述。组织切片均采用石蜡包埋,除少量特殊染色外,绝大部分使用HE染色。

本书既可作为农业高等院校的动物医学与动物生产类(包括兽医、动物科学、牧医师资、试点专业)等学生的参考用书,也可作为农牧业院校和生物科学等专业的师生教学用书及供畜牧兽医科技人员参考。

作者简介

苏布登格日勒，蒙古族，内蒙古农业大学兽医学院副教授，硕士研究生导师，主要从事动物组织胚胎与发育生物学的教学及研究。主编及参编教材5部，主编专著1部，在相关核心期刊发表论文10余篇。

作者简介

李海军，理学博士，内蒙古农业大学兽医学院教授，博士研究生导师，基础兽医学学科主任。中国畜牧兽医学会组织胚胎学分会理事，内蒙古解剖学会常务理事兼副秘书长，内蒙古生物工程学会常务理事。主要从事哺乳动物卵泡发生与胚胎发育机理研究。曾获内蒙古自然科学奖一等奖1项；授权发明专利与软件著作权3项；主编、参编著作8部；在 Human Reproduction、Biology of Reproduction、Reproduction 等期刊发表论文20余篇。

丛书编委会

骆驼精品图书出版工程

主任委员 何新天（中国畜牧业协会）
芒　来（内蒙古农业大学）
姚新奎（新疆农业大学）
刘强德（中国畜牧业协会）
主　编 吉日木图（内蒙古农业大学）
副主编 阿扎提·祖力皮卡尔（新疆畜牧科学院）
哈斯苏荣（内蒙古农业大学）
委　员 双　全（内蒙古农业大学）
何飞鸿（内蒙古农业大学）
娜仁花（内蒙古农业大学）
苏布登格日勒（内蒙古农业大学）
那仁巴图（内蒙古农业大学）
明　亮（内蒙古农业大学）
伊　丽（内蒙古农业大学）
周俊文（内蒙古自治区阿拉善盟畜牧研究所）
张文彬（内蒙古自治区阿拉善盟畜牧研究所）
斯仁达来（内蒙古农业大学）
郭富城（内蒙古农业大学）
马萨日娜（内蒙古农业大学）
海　勒（内蒙古农业大学）
好斯毕力格（内蒙古戈壁红驼生物科技有限责任公司）
王文龙（内蒙古农业大学）
嘎利兵嘎（内蒙古农业大学）
李海军（内蒙古农业大学）
任　宏（内蒙古农业大学）
道勒玛（内蒙古自治区阿拉善盟畜牧研究所）
额尔德木图（内蒙古自治区锡林郭勒盟苏尼特右旗畜牧兽医工作站）

前 言

骆驼主要分布于我国西北和华北的荒漠与半荒漠地区，是这一地区草原畜牧业的重要组成部分。千百年来，骆驼与良好适应环境的其他家畜如绵羊与山羊等一起共同构成了这一地区广泛而独特的畜牧业资源体系，对于因地制宜地发展边疆畜牧业生产具有十分重要的意义。目前有关各种家畜正常器官组织结构的参考书多已问世，但是系统介绍骆驼器官组织学结构的书籍尚未见到。我们编著《骆驼组织学》的目的就是搭建骆驼器官组织学结构体系，以填补国内与国际相关骆驼研究空白。

本书详细描述了骆驼各系统器官组织学结构，具有以下特点：第一，内容精炼而系统。在编排方面，按常规组织学的学习规律，根据系统顺序分别介绍低倍与高倍视野下的整体与局部结构；在内容选择上，精选了骆驼各主要代表性器官的组织切片；在内容表现上，除了在图片上直接标注外，各章前还有一段精炼的有关该系统组织结构的文字概述；书后附有常见组织学名词中英文对照。第二，图片质量佳。书中所有的显微照片均为原创。新鲜标本材料来源于屠宰厂。图片逼真清晰，是一本高质量的骆驼组织学彩色图谱。第三，实用性强。组织切片均采用石蜡包埋，除少量特殊染色外，绝大部分使用HE染色，更具实用性。

由于编著者水平有限，书中难免有错漏之处，敬请读者批评指正。

著 者

2020年11月1日

FOREWORD

Camel is mainly distributed in the desert and semi-desert areas in China. For thousands of years, camel, together with other well adapted animals, such as sheep and goats, have formed the extensive and unique livestock resource system, which is of great significance to livestock development in these border areas. Many reference books on the normal pattern of various livestock histology have been published, but those about camel's have not yet been seen. The present book has the ambition to contribute to this knowledge.

This book covers microscopic description of the normal pattern of the camel histology. The whole and the local structure under the low and the high vision are introduced according to the system order; the main representative organs of camel are selected for the preparation of the tissue sections; in addition to the picture directly marked, there is a brief overview about the histological pattern in the beginning of every chapter. Secondly, all microphotographs in this book are original. Fresh specimens come from slaughterhouses. The book is a high quality histological color map of camel. Thirdly, with the exception of a few special stains, most are stained with H-E method, which is the most common and practical stain procedure in most laboratories in China.

Although we have made every effort to present the work in its best, omissions and errors could be expected. Accordingly, any suggestions and comments about the atlas would be highly appreciated and could help us in improving the coming edition.

Author
November 1, 2020

目 录

前言

第一章　神经系统 ·· 1

第二章　心血管系统 ·· 15

第三章　免疫系统 ·· 29

第四章　内分泌系统 ·· 43

第五章　消化道 ·· 59

第六章　消化腺 ·· 97

第七章　呼吸系统 ·· 113

第八章　泌尿系统 ·· 123

第九章　雄性生殖系统 ·· 135

第十章　雌性生殖系统 ·· 153

第十一章　被皮系统与骨骼肌 ·· 165

CONTENTS

FORWARD

CHAPTER 1 Nervous System ··· 2

CHAPTER 2 Cardiovascular System ·· 16

CHAPTER 3 Cardiovascular System ·· 30

CHAPTER 4 Endocrine System ··· 44

CHAPTER 5 Digestive Tract ·· 60

CHAPTER 6 Digestive Glands ··· 98

CHAPTER 7 Respiratory System ··· 114

CHAPTER 8 Urinary System ··· 124

CHAPTER 9 Male Reproductive System ·· 136

CHAPTER 10 Female Reproductive System ··· 154

CHAPTER 11 Integumentary System and Skeletal Muscles ···················· 166

第一章 神经系统

CHAPTER 1

　　骆驼神经系统是骆驼机体内重要的调节系统，支配和调节全身各组织器官的活动。骆驼神经系统由大脑、小脑、脊髓和与之相连的神经共同组成。神经系统与内分泌和免疫系统密切配合，形成神经—免疫—内分泌网络，共同调节机体的生理活动。

CHAPTER 1

Nervous System

The nervous system of camel contains the cerebrum, cerebellum, spinal cord, and nerves connecting them to each other, detects, analyzes, and possibly uses and transmits all information generated by sensory stimuli including heat, light, mechanical, electrical, and chemical changes that occur externally and internally. And it organizes, integrates, and coordinates different functions of the body as a whole, including motor, visceral, endocrine, and mental activities.

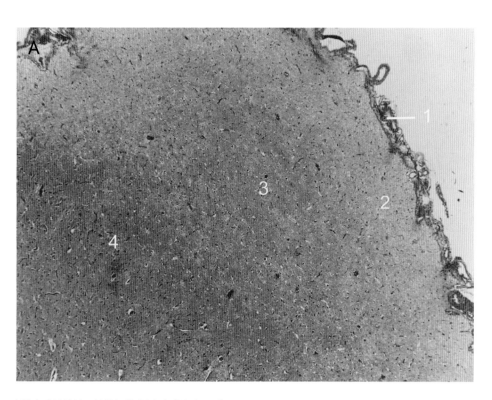

A.大脑皮质的外层，神经细胞分层不明显（40×）。
A. The outer layer of the cerebral cortex, the stratification of nerve cells is not obvious (40×).

1.软脑膜	1. Pia mater
2.分子层	2. Molecular layer
3.外颗粒层	3. External granular layer
4.锥体细胞层	4. Pyramidal layer

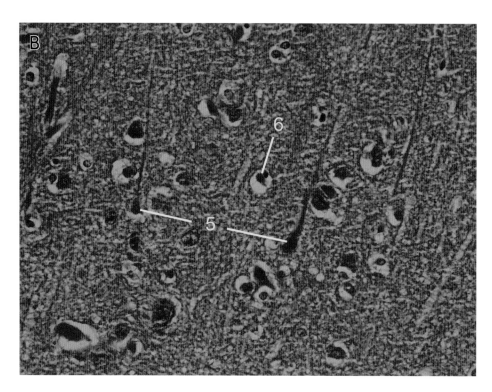

B.锥体细胞，高倍观察（400×）。
B. Pyramidal cell, high magnification observation of the pyramidal cell（400×）.

5.锥体细胞	5. Pyramidal cell
6.毛细血管	6. Capillary

图 1-1　大脑

Fig.1-1　Cerebrum

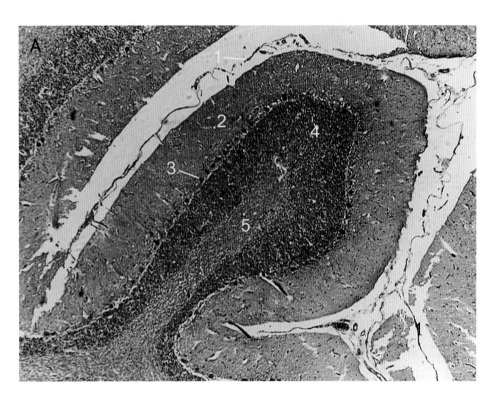

A.小脑分层，小脑脑回和脑沟很明显（40×）。
A. Cerebellar stratification, the gyrus and sulcus are clear (40×).

1.软脑膜	1. Pia mater
2.分子层	2. Molecular layer
3.浦肯野细胞层	3. Purkinje cell layer
4.颗粒层	4. Granular layer
5.髓质	5. Medulla

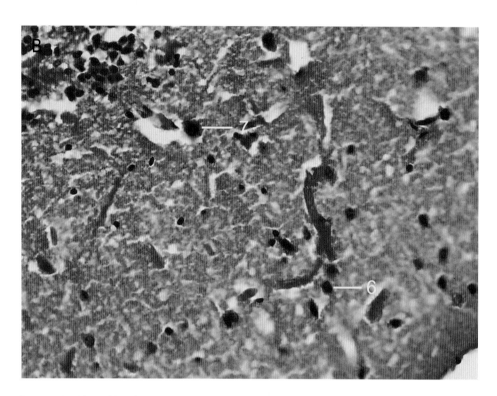

B. 分子层，此层神经元较少（400×）。
B. Molecular layer, less neurons (400×).

6. 星形细胞	6. Stellate cell
7. 篮状细胞	7. Basket cell

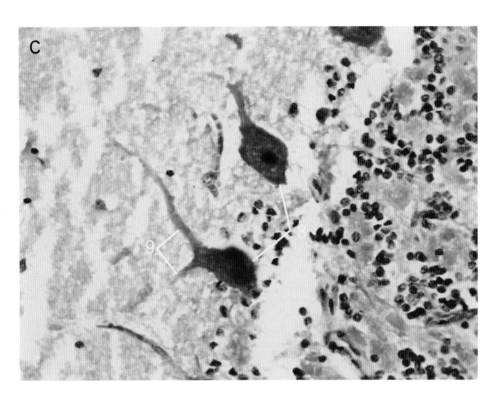

C.浦肯野细胞，胞体呈梨形，从胞体顶端发出2～3个树突伸入分子层（400×）。
C. Purkinje cell, pear-shaped, sending 2-3 dendrites from its tip into the molecular layer（400×）.

8.浦肯野细胞	8. Purkinje cell
9.浦肯野细胞树突	9. Purkinje cell dendrite

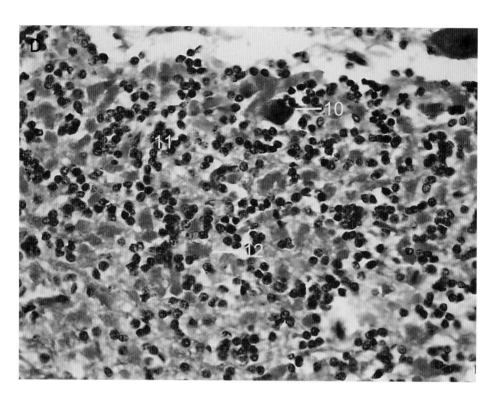

D. 颗粒层，神经细胞分布密集（400×）。
D. Granular layer, neurons are densely distributed (400×).

10. 高尔基细胞	10. Golgi cell
11. 颗粒细胞	11. Granular cell
12. 小脑小球	12. Cerebellar glomerulus

图1-2 小脑

Fig.1-2 Cerebellum

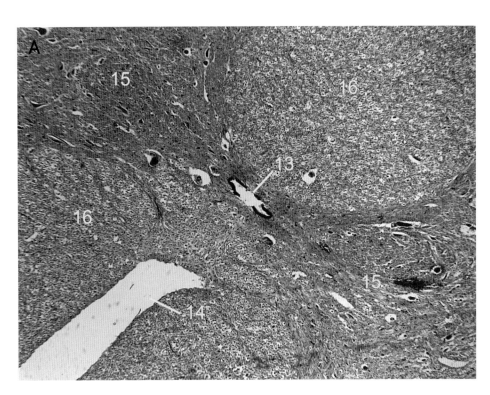

A.脊髓灰质及白质，灰质位于中央呈H形，周围是白质（40×）。
A. The gray and white matter, the H-shaped gray matter located in the central, surrounded by the white matter （40×）.

13.中央管	13. Central canal
14.腹正中裂	14. Ventral median fissure
15.灰质	15. Gray matter
16.白质	16. White matter

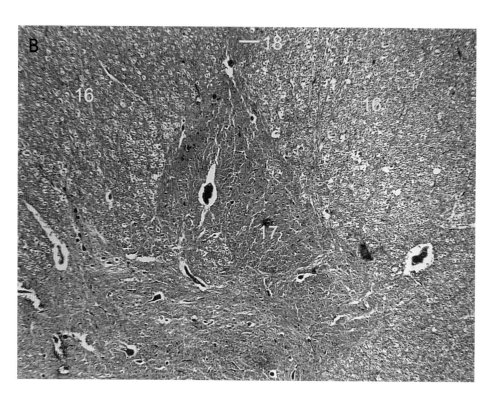

B.灰质背角，神经元分布较多（40×）。
B. The dorsal horn, neurons are densely distributed (40×).

16.白质	16. White matter
17.背角	17. Dorsal horn
18.背外侧束	18. Back lateral

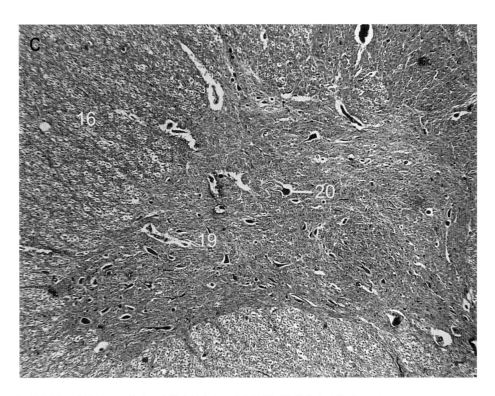

C. 灰质侧腹角，侧角神经元较少，胞体中等大小，腹角神经元胞体大小不等（40×）。
C. The lateral and ventral horns, less and medium sized neurons in the lateral horn, different sizes neurons in the ventral horn (40×).

16. 白质	16. White matter
19. 腹角	19. Ventral horn
20. 侧角神经元	20. Neurons in dorsal horn

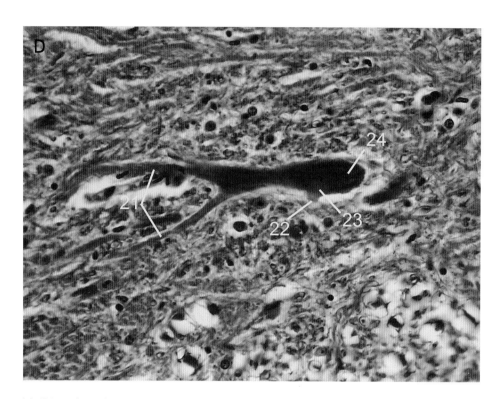

D. 腹角神经元（400×）。
D. Neurons in the ventral horn (400×).

21. 多极神经元树突	21. Multipolar neuron dendrites
22. 多极神经元轴突	22. Multipolar neuron axons
23. 轴丘	23. Axon hillock
24. 尼氏体	24. Nissl bodies

图 1-3　脊髓

Fig.1-3 Spinal cord

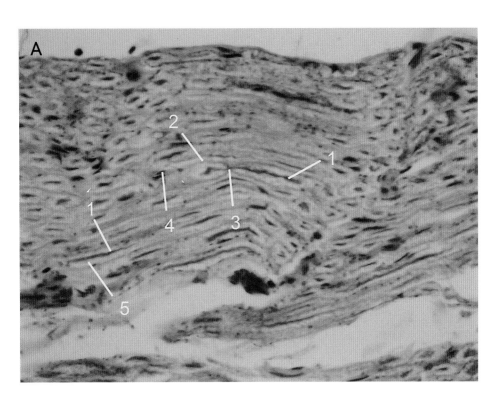

A.有髓神经纤维（400×）。
A. Myelinated nerve fiber （400×）.

1.轴索	1. Axial cord
2.神经膜	2. Schwann cell membrane
3.郎飞结	3. Ranvier node
4.施万细胞核	4. Schwann cell nuclei
5.髓鞘	5. Myelin sheath

图1-4 坐骨神经

Fig.1-4 Sciatic nerve

第二章 心血管系统

CHAPTER 2

骆驼心血管系统由心脏、动脉、毛细血管和静脉组成，是一个分支的封闭管道系统。心脏是促使血液流动的动力泵，其经过收缩与舒张，可将血液输入动脉。动脉经各级分支将血液输送到毛细血管。静脉由毛细血管汇合移行而来，将经过物质交换的血液回流至心脏。

CHAPTER 2

Cardiovascular System

The cardiovascular system of camel includes the heart, arteries, veins and capillaries. The heart pumps blood and the blood vessels channel and deliver it throughout the body. Arteries carry blood filled with nutrients away from the heart to all parts of the body. Capillaries join together to form veins which deliver deoxygenated blood back to the heart.

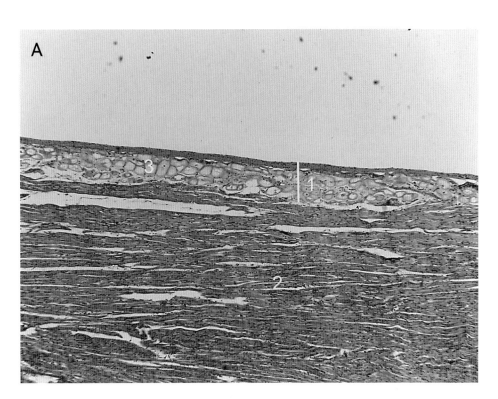

A.心壁心内膜及部分心肌膜，在心内膜中含有大量的浦肯野纤维（40×）。
A. The endocardium and myocardium, many Purkinje fibers in the endocardium （40×）.

1.心内膜	1. Endocardium
2.心肌膜	2. Myocardium
3.浦肯野纤维	3. Purkinje fiber

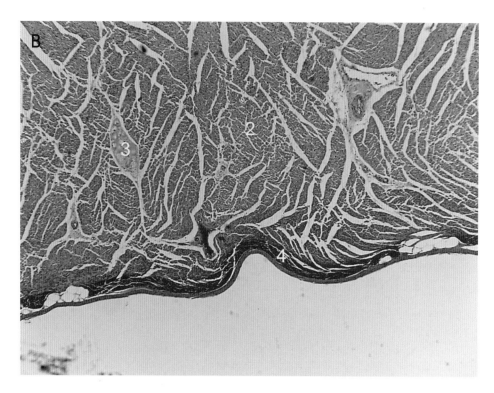

B.部分心肌膜与心外膜，心外膜属于浆膜（40×）。
B. The myocardium and epicardium, the latter belongs to serous membrane （40×）.

2.心肌膜	2. Myocardium
3.浦肯野纤维	3. Purkinje fiber
4.心外膜	4. Epicardium

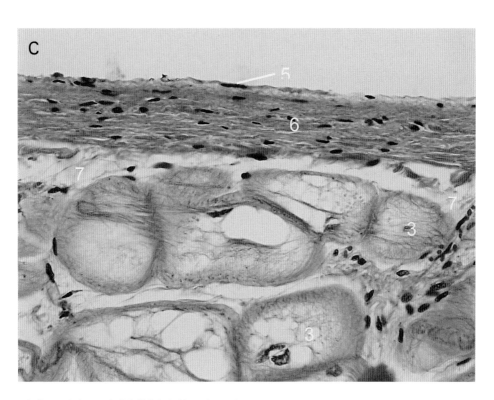

C.心内膜，心内皮下层为致密结缔组织较厚（400×）。
C. The endocardium, the subendothelial layer belongs to the dense connective tissue （400×）.

3.浦肯野纤维	3. Purkinje fiber
5.心内皮	5. Cardiac endothelium
6.心内皮下层	6. Subendothelial layer
7.心内膜下层	7. Subendocardial layer

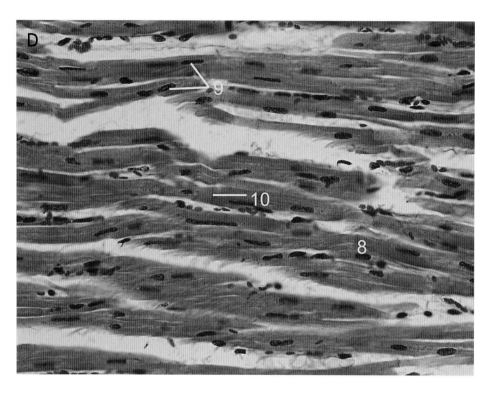

D. 纵切心肌细胞，心肌细胞粗长，闰盘清楚（400×）。
D. Cardiac muscle in longitudinal section, cardiac muscle fibers long and thick, intercalated disks clear （400×）.

8. 心肌纵切	8. Cardiac muscle in longitudinal section
9. 心肌细胞核	9. Nuclear of cardiac muscle cell
10. 闰盘	10. Intercalated disk

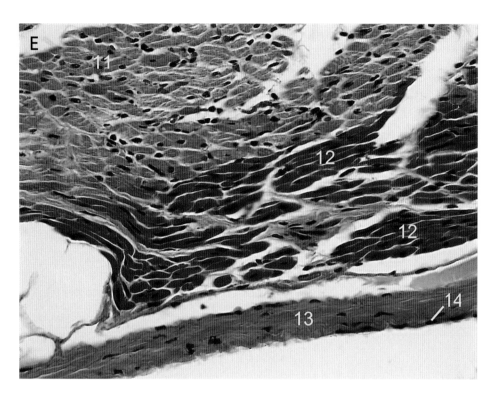

E.心肌不同切面及心外膜，心外膜表面为间皮（400×）。
E. The different section views of cardiac muscle and epicardium, mesothelium located on its surface (400×).

11. 心肌横切	11. Cardiac muscle in cross section
12. 心肌斜切	12. Cardiac muscle in bevel section
13. 结缔组织	13. Connective tissue
14. 间皮	14. Mesothelium
15. 脂肪细胞	15. Adipocyte

图2-1 心脏
Fig.2-1 Heart

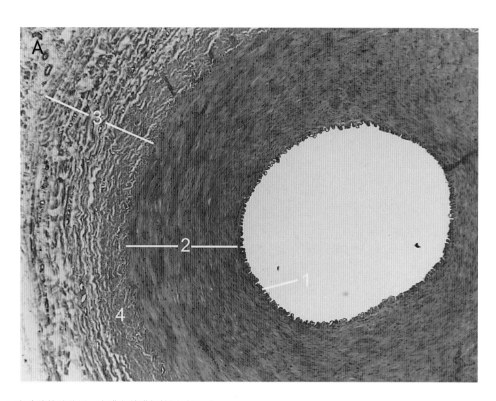

A.中动脉管壁分层，中膜和外膜都较厚（40×）。
A. The medium artery wall stratification, the middle and external tunics are thick (40×).

1.内膜	1. Internal tunic
2.中膜	2. Middle tunic
3.外膜	3. External tunic
4.外弹性膜	4. External elastic membrane

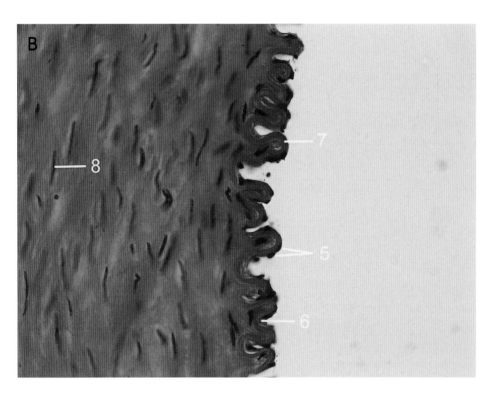

B. 中动脉内膜及中膜平滑肌，内皮下层较薄，内弹性膜明显（400×）。
B. The internal tunic stratification of medium artery, the thin subendothelial layer and distinct internal elastic membrane (400×).

5. 内皮	5. Endothelium
6. 内皮下层	6. Subendothelial layer
7. 内弹性膜	7. Internal elastic membrane
8. 平滑肌细胞	8. Smooth muscle cell

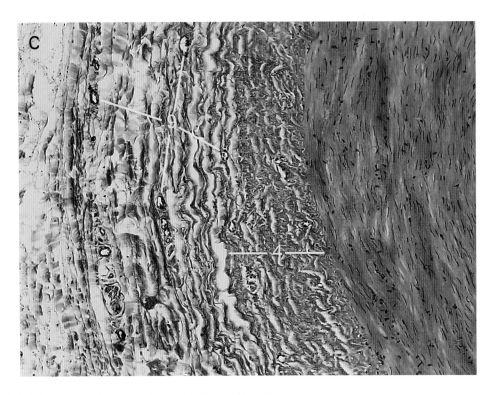

C.中动脉外弹性膜及外纤维膜，外膜富含滋养血管（100×）。
C. The external elastic and external fibrous membranes of medium artery, plenty of nutrient vessels in the external tunic (100×).

4.外弹性膜	4. External elastic membrane
9.营养血管	9. Nutrient vessels

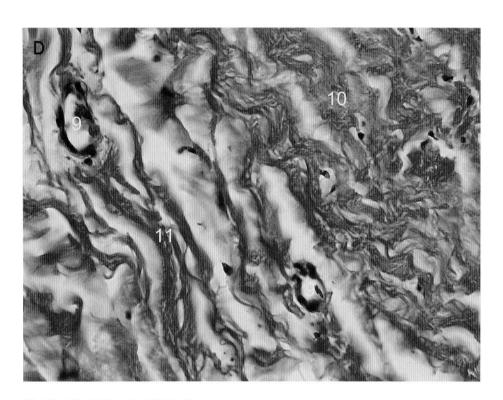

D.弹性纤维及胶原纤维，弹性纤维粗而长（400×）。
D. The elastic and collagenous fibers, the elastic fibers are thick and long （400×）.

9.营养血管	9. Nutrient vessels
10.弹性纤维	10. Elastic fiber
11.胶原纤维	11. Collagenous fiber

图2-2　中动脉

Fig.2-2　Medium artery

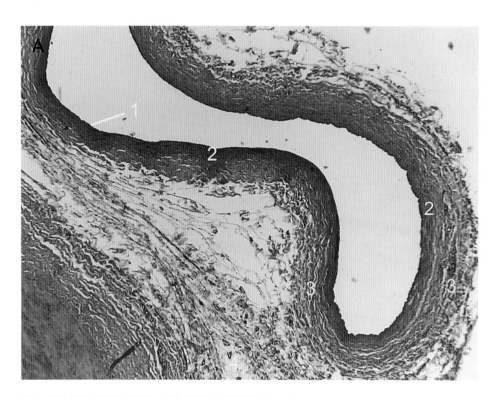

A.中静脉管壁分层，管腔不规则，内膜不明显，外膜稍厚（40×）。
A. Medium vein stratification, the irregular lumen and unclear internal tunic (40×).

1.内膜	1. Internal tunic
2.中膜	2. Middle tunic
3.外膜	3. External tunic

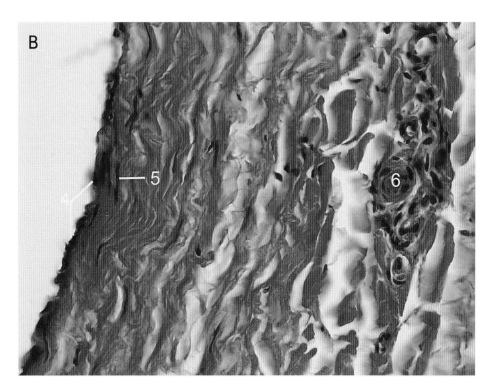

B.中静脉管壁，缺乏内弹性膜，外膜不含弹性纤维，富含滋养血管（400×）。
B. The internal elastic membrane is absent in medium vein, less elastic fibers and more nutrient vessels in the external tunic（400×）.

4.内皮	4. Endothelium
5.平滑肌细胞	5. Smooth muscle cell
6.营养血管	6. Nutrient vessels

图2-3　中静脉

Fig.2-3　Medium vein

第三章 免疫系统

CHAPTER 3

作为骆驼机体防御屏障，免疫系统包含了淋巴细胞产生与循环的所有相关结构，如脾脏、胸腺、骨髓等。淋巴组织存在于多种免疫器官中（如淋巴结），以及与消化系统相连接的淋巴滤泡中（如扁桃体）。

CHAPTER 3

Cardiovascular System

The immune system of camel as the defense barrier of the body includes all the structures dedicated to the circulation and production of lymphocytes, which consists of the spleen, thymus, bone marrow, and so on. Lymphoid tissue is found in many organs, particularly the lymph nodes, and in the lymphoid follicles associated with the digestive system such as tonsils.

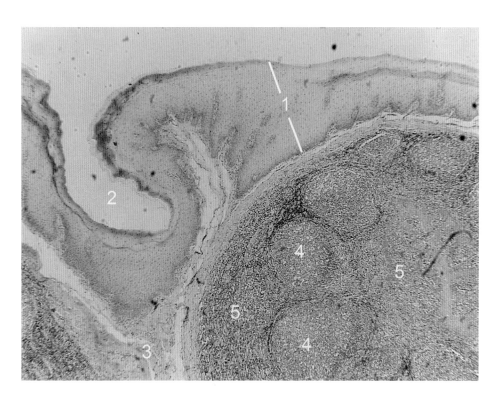

表面复层扁平上皮下陷形成隐窝（40×）。
The stratified squamous epithelium sagging to form the crypts (40×).

1.复层扁平上皮	1. Stratified squamous epithelium
2.隐窝	2. Crypt
3.结缔组织	3. Connective tissue
4.淋巴小结	4. Lymphoid nodule
5.弥散淋巴组织	5. Diffuse lymphoid tissue

图 3-1　腭扁桃体
Fig.3-1　Palatine tonsil

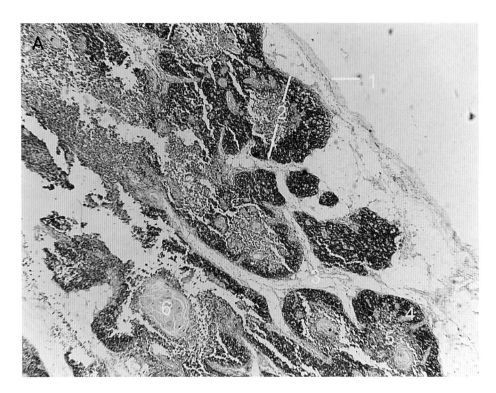

A.被膜较厚，由疏松结缔组织组成，髓质之间的小叶相连接（40×）。
A. The thick capsule consists of the loose connective tissue (40×).

1. 被膜	1. Capsule
2. 胸腺小叶	2. Thymic lobule
3. 小叶间隔	3. Interlobular septa
4. 胸腺皮质	4. Thymic cortex
5. 胸腺髓质	5. Thymic medulla
6. 胸腺小体	6. Thymic corpuscle

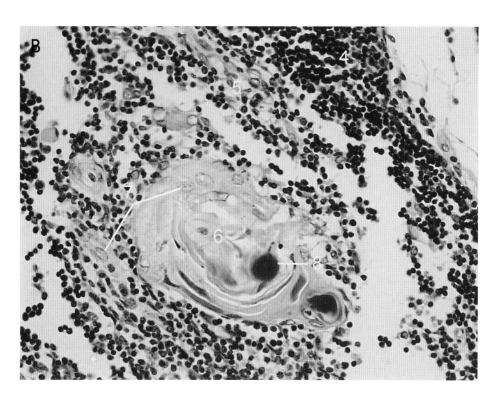

B.皮质和髓质部的一部分。髓质含有膨大而角质化的胸腺上皮细胞形成胸腺小体（400×）。
B. A part of its cortex and medulla, thymic corpuscles exist in the medulla (400×).

4.胸腺皮质	4. Thymic cortex
5.胸腺髓质	5. Thymic medulla
6.胸腺小体	6. Thymic corpuscle
7.胸腺上皮细胞	7. Thymic epithelial cell
8.完全角化的上皮细胞	8. The stratified squamous keratinized epithelial

图 3-2　胸腺

Fig.3-2　Thymus

A.脾实质，分为红髓和白髓，小梁自被膜延伸到整个红髓（40×）。
A. The red and white medulla, the trabecula extending from the capsule into the whole red medulla (40×).

1.被膜	1. Capsule
2.小梁	2. Trabecula
3.白髓	3. White pulp
4.红髓	4. Red pulp

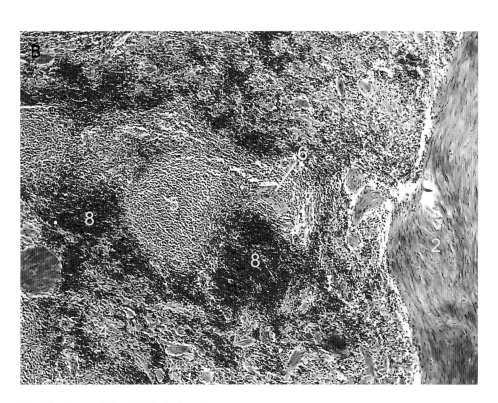

B.脾实质的一部分，边缘区比较发达（100×）。
B. Part of spleen with the developed marginal zone (100×).

2.小梁	2. Trabecula
4.红髓	4. Red pulp
5.淋巴小结	5. Lymphoid nodule
6.中央动脉	6. Central artery
7.动脉周围淋巴鞘	7. Periarterial lymphatic sheath
8.边缘区	8. Marginal zone

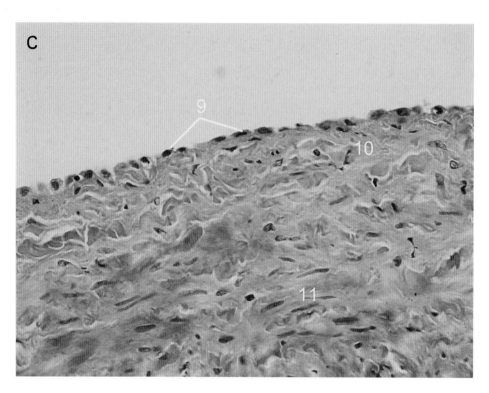

C.被膜厚,为致密结缔组织与间皮结构(400×)。
C. The capsule consists of the dense connective tissue and mesothelium (400×).

9.间皮	9. Mesothelium
10.结缔组织	10. Connective tissue
11.平滑肌	11. Smooth muscle

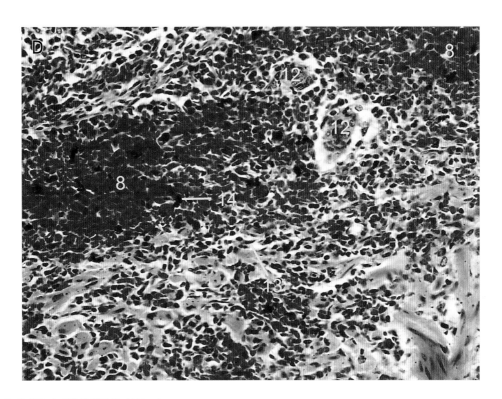

D.边缘区中有脾的椭圆体（400×）。
D. Nissl bodies exist in marginal zone （400×）.

8.边缘区	8. Marginal zone
12.椭圆体	12. Nissl bodies
13.内皮细胞	13. Endothelial cell
14.吞噬衰老红细胞的巨噬细胞	14. Macrophage swallowing the aging erythrocytes

图 3-3　脾脏
Fig.3-3 Spleen

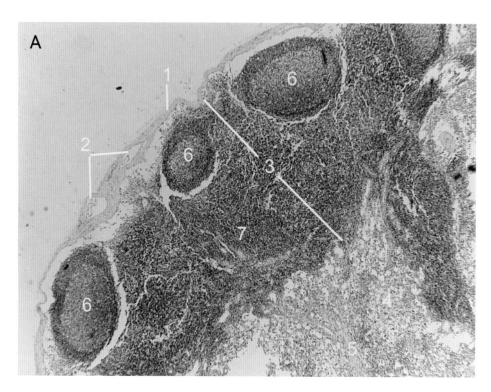

A.实质分为皮质和髓质，皮质中淋巴小结明显（40×）。
A. The cortex and medulla, the distinct lymphoid nodules in the cortex (40×).

1.被膜	1. Capsule
2.输入淋巴管	2. Afferent lymphatic vessel
3.皮质	3. Cortex
4.髓质	4. Medulla
5.小梁	5. Trabecula
6.淋巴小结	6. Lymphoid nodule
7.副皮质区	7. Paracortex zone

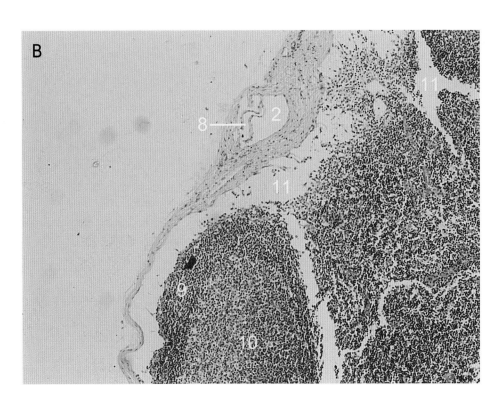

B. 被膜上清晰可见输入淋巴管瓣膜（100×）。
B. The valve of afferent lymphatic vessel is clear in the capsule （100×）.

2. 输入淋巴管	2. Afferent lymphatic vessel
8. 瓣膜	8. Valve
9. 小结帽	9. Nodule cap
10. 生发中心	10. Germinal center
11. 皮质淋巴窦	11. Cortical sinus

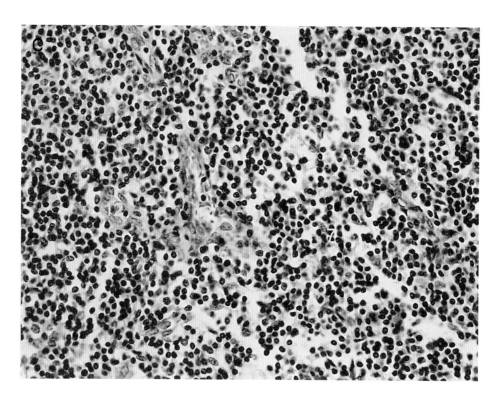

C. 淋巴结副皮质区中毛细血管后微静脉（400×）。
C. The post capillary venules in paracortex zone of Lymph node (400×).

7. 副皮质区	7. Paracortex zone
12. 毛细血管后微静脉（横切与纵切面）	12. Post capillary venule

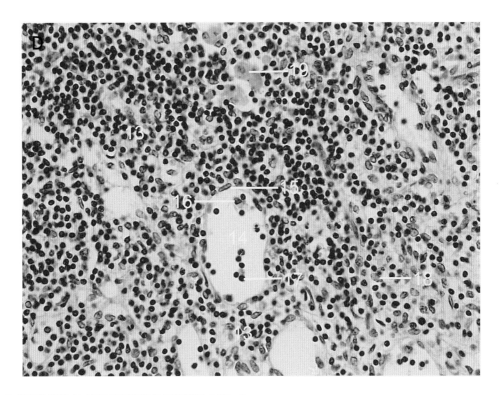

D. 淋巴结髓质中可见吞噬红细胞的巨噬细胞（400×）。
D. The macrophages swallowing the erythrocytes in lymph node medulla （400×）.

13. 髓索	13. Medullary cord
14. 髓质淋巴窦	14. Medullary sinus
15. 内皮细胞	15. Endothelial cell
16. 巨噬细胞	16. Macrophage
17. 淋巴细胞	17. Lymphocyte
18. 网状细胞	18. Reticular cell
19. 吞噬红细胞的巨噬细胞	19. Macrophage swallowing the erythrocytes

图 3-4　淋巴结

Fig.3-4 Lymph node

第四章 内分泌系统

CHAPTER 4

骆驼内分泌系统由一系列内分泌腺组成，这些内分泌腺将激素与化学信号直接分泌至血液，并调节机体活动。内分泌腺无导管，毛细血管丰富，激素储存于胞内滤泡或分泌颗粒中。骆驼的内分泌器官结构，如肾上腺、脑垂体、甲状腺与松果体等，与其他哺乳动物并无二致。

CHAPTER 4

Endocrine System

The endocrine system of camel is a system of glands secreting hormone or chemicals directly into the bloodstream to regulate the body. Features of endocrine glands are, in general, their ductless nature, their vascularity, and usually the presence of intracellular vacuoles or granules storing their hormones. Pure endocrine glands of the camel similar to other mammals include adrenals, pituitary, thyroid and pineal.

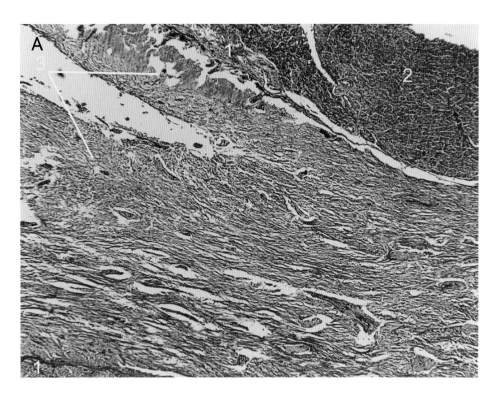

A.垂体漏斗部、结节部和远侧部（40×）。
A. Infundibularis partes, pars tuberalis and pars distalis in hypophysis （40×）.

1.结节部	1. Pars tuberalis
2.远侧部	2. Pars distalis
3.漏斗部	3. Infundibularis partes

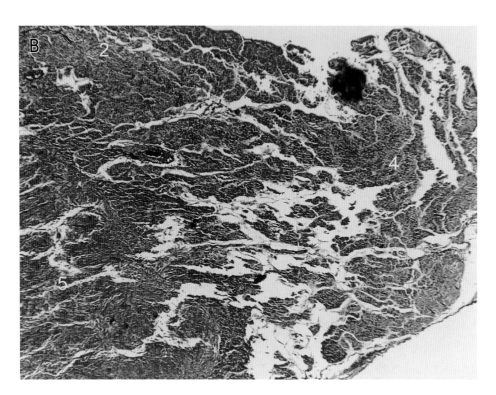

B.垂体中间部和神经部（40×）。
B. Pars intermedia and pars neuralis in hypophysis (40×).

2.远侧部	2. Pars distalis
4.中间部	4. Pars intermedia
5.神经部	5. Pars neuralis

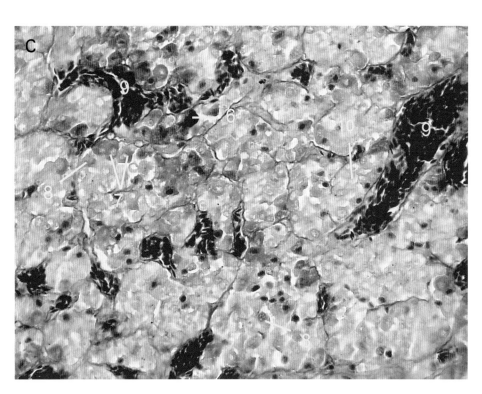

C. 远侧部结构，切片上催乳激素细胞数量少（400×）。
C. The pars distalis, less mammotroph cells (400×).

6. 催乳激素细胞	6. Mammotroph cell
7. 生长激素细胞	7. Somatotroph cell
8. 嗜碱性细胞	8. Basophilic cell
9. 血窦	9. Sinus
10. 嫌色细胞	10. Chromophobe cell

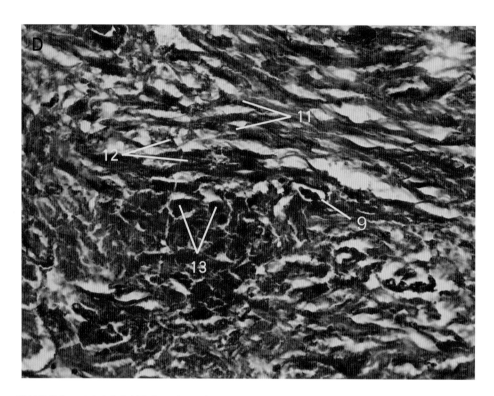

D. 神经部结构，具有独特的纤维外观（400×）。
D. The pars neuralis with the special nerve fibers (400×).

9. 血窦	9. Sinus
11. 无髓神经纤维	11. Unmyelinetednerve cell
12. 垂体细胞	12. Hypophysis cell
13. 赫令小体	13. Herring body

图 4-1 垂体

Fig.4-1 Hypophysis

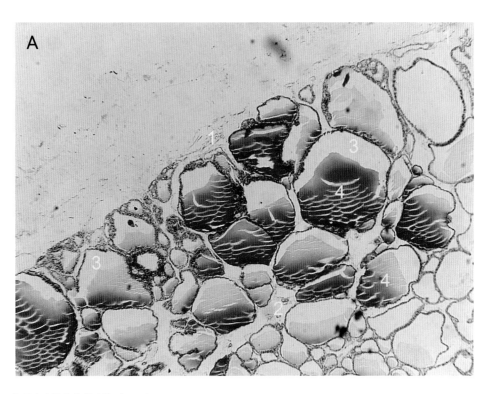

A.大量内含胶状物的滤泡（40×）。
A. Many follicles containing glue (40×).

1.被膜	1. Capsule
2.毛细血管	2. Capillary
3.滤泡	3. Follicle
4.胶状物	4. Glue

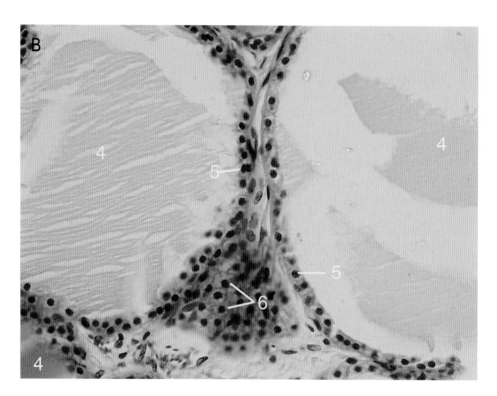

B.滤泡旁细胞大，淡染（400×）。
B. The parafollicular cells are large and light dyeing （400×）.

4.胶状物	4. Glue
5.滤泡上皮细胞	5. Follicular endothelial cell
6.滤泡旁细胞	6. Parafollicular cell

图4-2 甲状腺

Fig.4-2 Thyroid gland

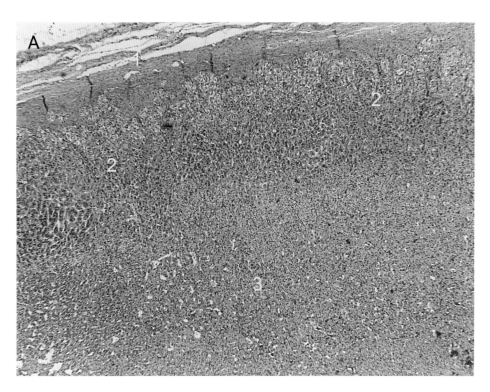

A.肾上腺皮质和被膜，多形带细胞排列不规则（40×）。
A. The capsule and cortex, irregular cell arranged in the zone multiformis (40×).

1.被膜	1. Capsule
2.多形带	2. Zona multiformis
3.束状带	3. Zona fasciculata

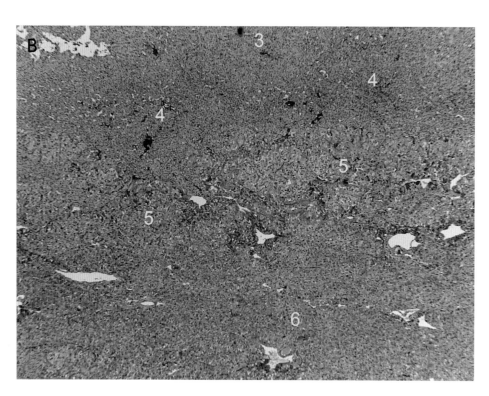

B. 肾上腺髓质与网状带，髓质内外区细胞染色对比鲜明（40×）。
B. The zone reticularis and medulla, showing the different staining view of cells in medullary inner and outer regions (40×).

3. 束状带	3. Zona fasciculata
4. 网状带	4. Zona reticularis
5. 髓质外区	5. Medullary outer region
6. 髓质内区	6. Medullary inner region

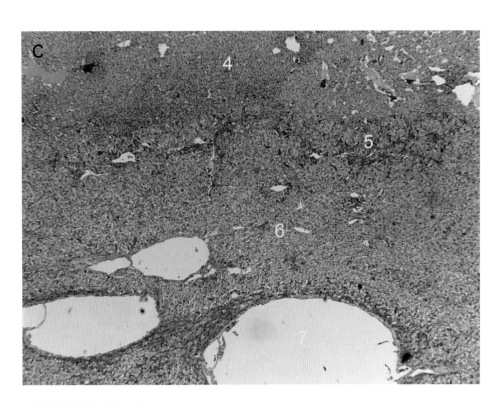

C.髓质中央有中央静脉（40×）。
C. The central vein located in the medulla (40×).

4.网状带	4. Zona reticularis
5.髓质外区	5. Medullary outer region
6.髓质内区	6. Medullary inner region
7.中央静脉	7. Central vein

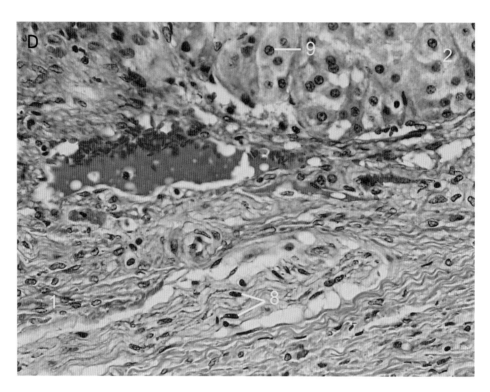

D.被膜与皮质部分结构。被膜内有上皮样细胞（400×）。
D. The capsule and part of the cortex, epithelioid cells exist in the capsule (400×).

1. 被膜	1. Capsule
2. 多形带	2. Zona multiformis
8. 上皮样细胞	8. Epithelial cell
9. 多形带细胞	9. Zona multiformis cell

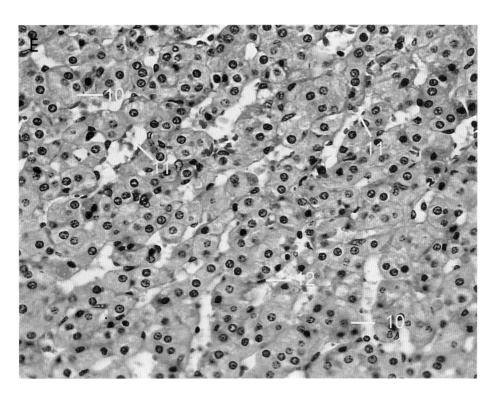

E. 皮质束状带，细胞体积大，成条束状排列（400×）。
E. The zona fasciculata, showing the large cells arranged in bundles (400×).

10. 束状带细胞	10. Zona fasciculata cell
11. 血窦	11. Sinus
12. 内皮细胞	12. Endothelial cell

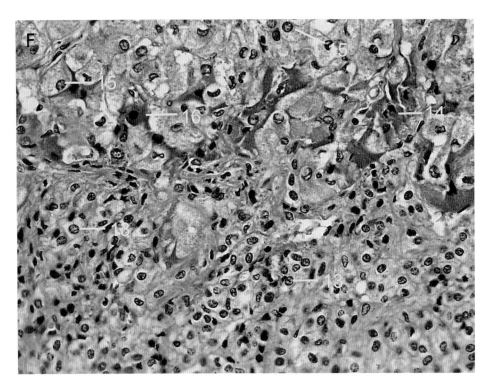

F.网状带与髓质，髓质细胞间可见神经节细胞（400×）。
F. The medulla and zona reticularis, showing the gangliocytes among the medullary cells （400×）.

13.网状带细胞	13. Zona reticularis cell
14.髓质外区细胞	14. Cell in medullary outer region
15.髓质内区细胞	15. Cell in medullary inner region
16.神经节细胞	16. Gangliocyte

图4-3 肾上腺

Fig.4-3 Adrenal gland

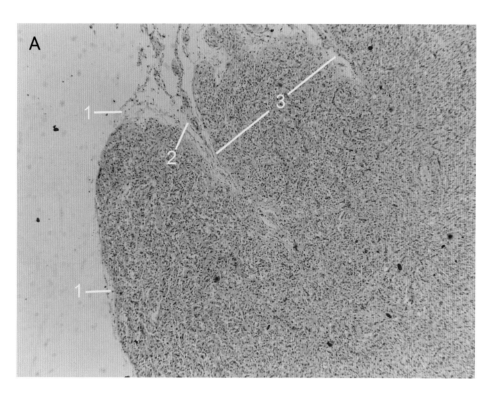

A.实质有若干不规则的小叶（40×）。
A. Several irregular lobules in the parenchyma (40×).

1.被膜	1. Capsule
2.小叶间隔	2. Interlobular septa
3.小叶	3. Lobule

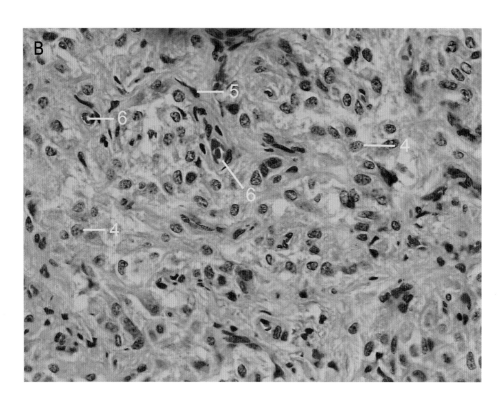

B.松果体结构（400×）。
B. Pineal composition（400×）.

4.松果体细胞	4. Pinealocyte
5.神经胶质细胞	5. Neuronglial cell
6.血窦	6. Sinus

图4-4 松果体
Fig.4-4 Pineal body

第五章 消化道

CHAPTER 5

骆驼消化道是一条衬有上皮并且粗细不等的连续管道，包括口腔、咽、食管、胃、小肠、大肠和肛门，其主要功能是摄取和消化食物、吸收食物中的营养并排出残渣。基于其特殊的结构特点，因此骆驼能够很好地适应低质量纤维食物的摄入。

CHAPTER 5

Digestive Tract

The digestive tract of camel is a continuous tube lined with epithelium and varying thickness, includes oral cavity, esophagus, stomach, small and large intestines, rectum and anus. The main functions of the digestive tract are to ingest and digest food, absorb nutrients from food and remove residues. By the structure and the function of their digestive tract, the camel is well adapted to intake low quality fibrous roughages.

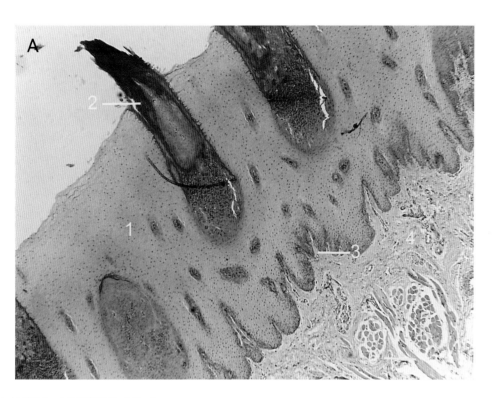

A. 丝状乳头,上皮细胞角化（40×）。
A. The filiform papilla, the keratinized epithelium （40×）.

1.角化复层扁平上皮	1. Stratified squamous keratinized epithelial
2.丝状乳头	2. Filiform papilla
3.小乳头	3. Small papilla
4.固有层	4. Lamina propria
5.骨骼肌横切	5. Skeletal muscle in cross section
6.骨骼肌纵切	6. Skeletal muscle in longitudinal section

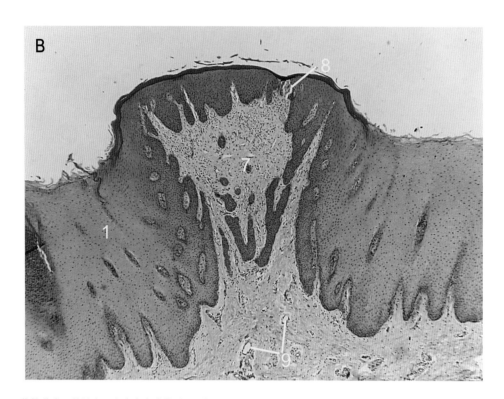

B.菌状乳头，数量少，上皮中有味蕾（40×）。
B. The fungiform papilla, less and several taste buds in its epithelium (40×).

1.角化复层扁平上皮	1. Stratified squamous keratinized epithelial
7.菌状乳头	7. Fungiform papilla
8.味蕾	8. Taste bud
9.小血管	9. Small blood vessels

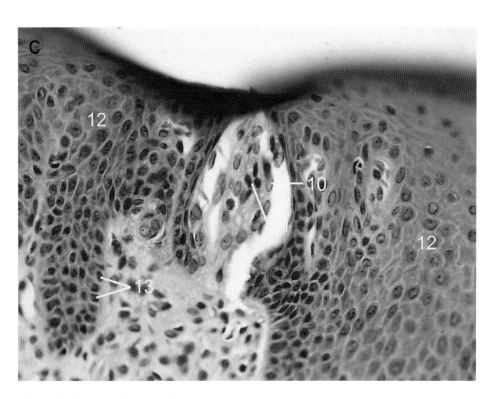

C. 味蕾，感觉细胞比支持细胞深染（400×）。
C. The taste buds, sensory cells dye deeper than supportive cells (400×).

10. 支持细胞	10. Supportive cell
11. 感觉细胞	11. Sensory cell
12. 棘层	12. Stratum spinosum
13. 基底层	13. Stratum basale

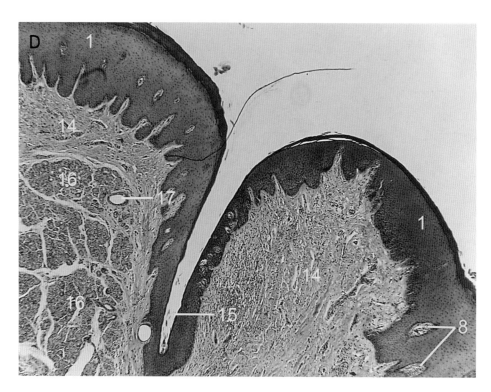

D.轮廓乳头，轮廓乳头位于上皮的凹陷内（400×）。
D. The circumvallate papilla, invaginated in the epithelium (400×).

1.角化复层扁平上皮	1. Stratified squamous keratinized epithelial
8.味蕾	8. Taste bud
14.轮廓乳头	14. Circumvallate papilla
15.环沟	15. Taste groove
16.唾液腺	16. Salivary glands
17.导管	17. Duct

图5-1 舌

Fig.5-1 Tongue

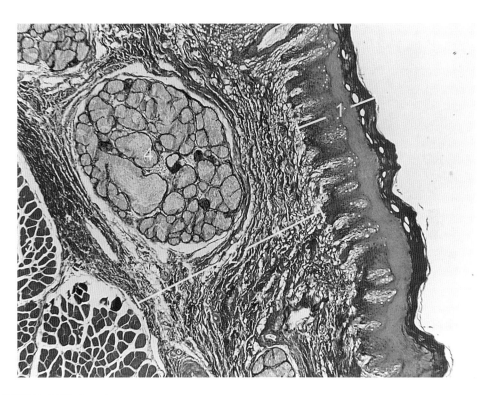

黏膜肌层不明显（40×）。
The muscularis mucosa is not obvious (40×).

1.角化复层扁平上皮	1. Stratified squamous keratinized epithelial
2.黏膜下层	2. Submucosa
3.肌层	3. Muscularis
4.食管腺	4. Esophageal gland

图5-2 食管前段

Fig. 5-2 Anterior segment of esophagus

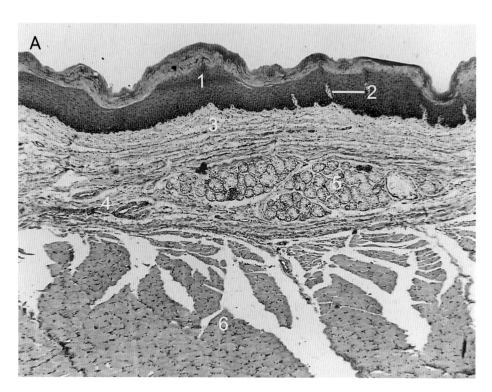

A.肌层为横纹肌，食管腺以黏液腺为主（40×）。
A. The cross-striated muscles, mainly mucous glands (40×).

1.角化复层扁平上皮	1. Stratified squamous keratinized epithelial
2.结缔组织乳头	2. Connective papilla
3.固有层	3. Lamina propria
4.黏膜下层	4. Submucosa
5.食管腺	5. Esophageal gland
6.内斜肌层	6. Internal oblique tunica muscularis

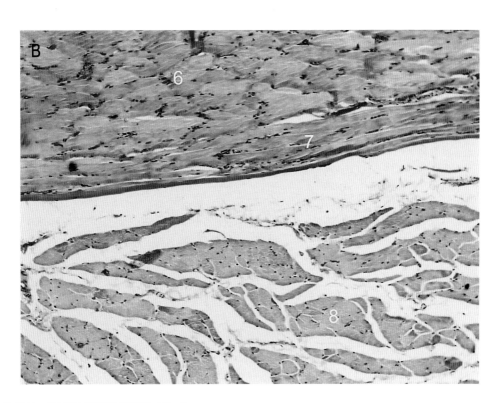

B.肌层，肌层分三层均为横纹肌（40×）。
B. The tunica muscularis, belong to cross-striated muscles (40×).

6.内斜肌层	6. Internal oblique tunica muscularis
7.中环肌层	7. Intermediate annular tunica muscularis
8.外纵肌层	8. External longitudinal tunica muscularis

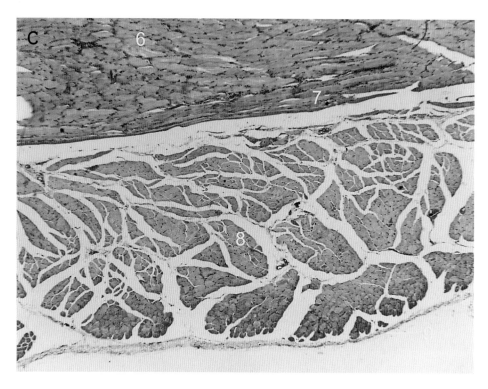

C. 外膜为浆膜（40×）。
C. The tunica adventitia is the serous membrane (40×).

6. 内斜肌层	6. Internal oblique tunica muscularis
7. 中环肌层	7. Intermediate annular tunica muscularis
8. 外纵肌层	8. External longitudinal tunica muscularis
9. 浆膜	9. Serous membrane

图5-3 食管后段

Fig.5-3 Posterior segment of esophagus

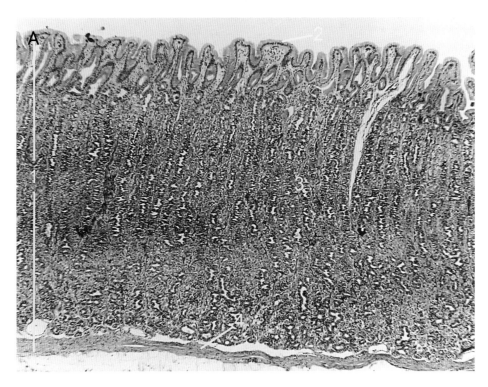

A.黏膜层，固有层很厚被大量胃腺所占据（40×）。
A. The mucous membrane, filled with glands （40×）.

1.黏膜	1. Mucous membrane
2.上皮	2. Epithelium
3.固有层	3. Lamina propria
4.黏膜肌层	4. Muscularis mucosa

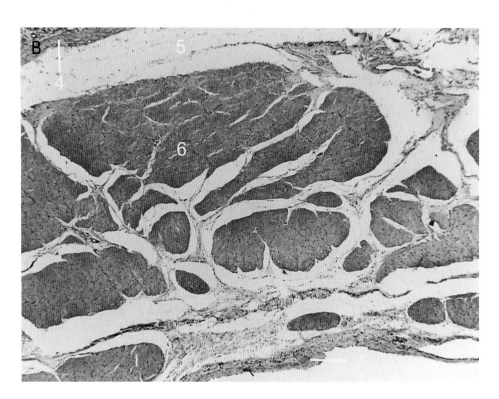

B.肌层发达，外膜为浆膜结构（40×）。
B. The developed tunica muscularis, tunica adventitia is serous membrane (40×).

4.黏膜肌层	4. Muscularis mucosa
5.黏膜下层	5. Submucosa
6.肌层	6. Tunica muscularis
7.外膜	7. Tunica adventitia

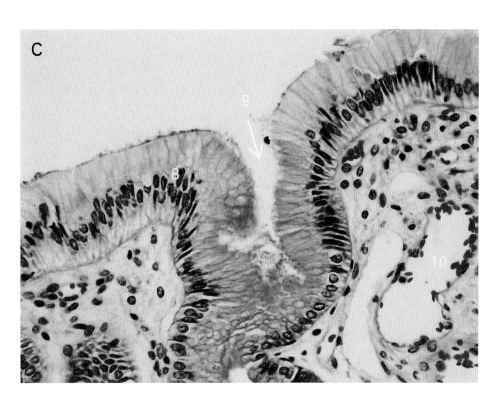

C.黏膜上皮，上皮为高柱状上皮淡染（400×）。
C. The mucous epithelium, high columnar epithelium with light dyeing (400×).

8.表层黏液细胞	8. Surface mucous cell
9.胃小凹	9. Gastric pit
10.固有层血管	10. Vessels in lamina propria

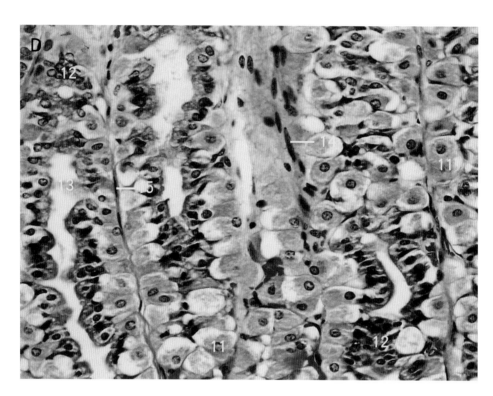

D. 胃底腺结构，主要由主细胞和壁细胞组成（400×）。
D. The fundic glands, mainly consists of the chief and parietal cells (400×).

11. 壁细胞	11. Parietal cell
12. 主细胞	12. Chief cell
13. 胃底腺腔	13. Fundic glandular antrum
14. 平滑肌细胞	14. Smooth muscle cell
15. 肌上皮细胞	15. Muscular epithelial cell

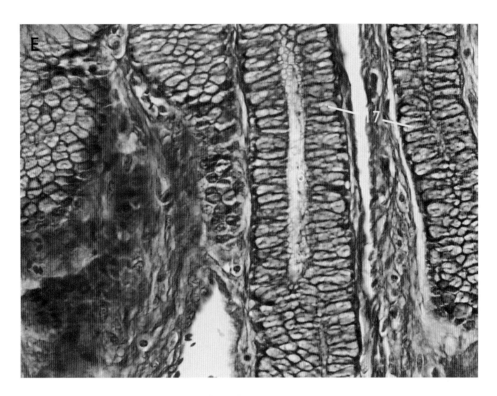

E.胃幽门腺，腺体管腔大，腺细胞为高柱状（400×）。
E. The pyloric glands, the glandular cavity is large and the glandular cells are high columnar (400×).

16.幽门腺腔	16. Pyloric glandular antrum
17.腺细胞	17. Glandular cell

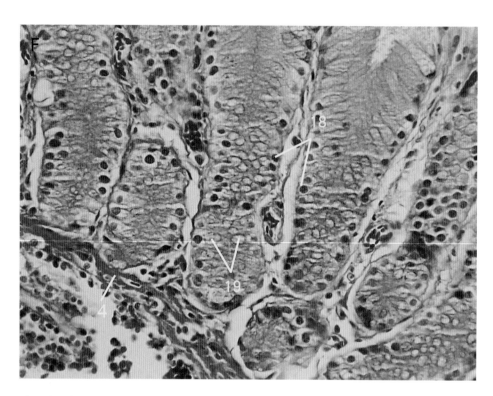

F. 胃贲门腺，腺上皮细胞内充满了黏原颗粒（400×）。
F. The cardiac glands, epithelial cell filled with mucin particles （400×）.

4. 黏膜肌层	4. Muscularis mucosa
18. 贲门腺细胞	18. Cardiac glandular cell
19. 黏原颗粒	19. Mucin particle

图5-4　胃

Fig.5-4 Stomach

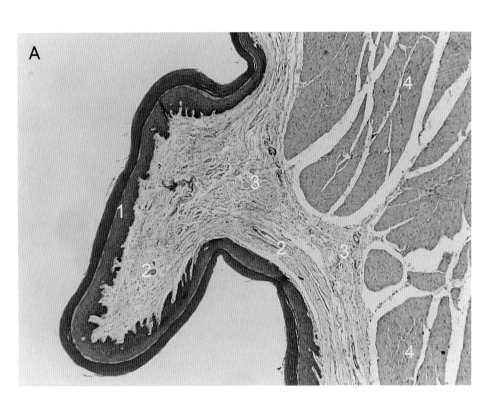

A.肌层，短桨状乳头（40×）。
A. The tunica muscularis, short oar papilla (40×).

1.角质化复层扁平上皮	1. Stratified squamous keratinized epithelial
2.固有层	2. Lamina propria
3.黏膜下层	3. Submucosa
4.肌层	4. Tunica muscularis

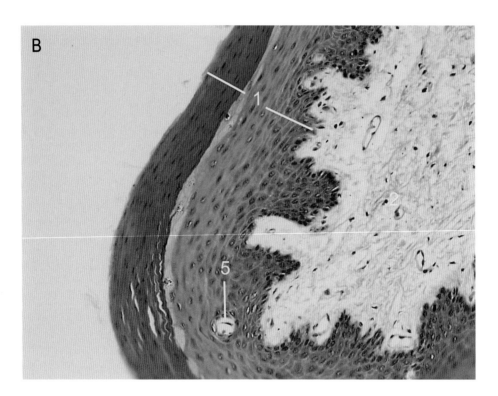

B.角质化上皮，上皮中含有毛细血管（200×）。
B. The keratinized epithelial, containing capillary (200×).

1.角质化复层扁平上皮	1. Stratified squamous keratinized epithelial
2.固有层	2. Lamina propria
5.上皮毛细血管	5. Capillary

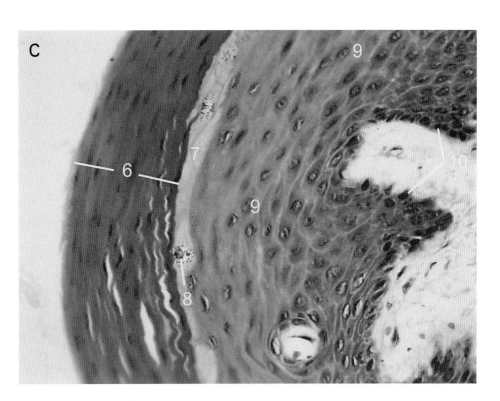

C.上皮角质化程度低（400×）。
C. Low keratinized epithelium（400×）.

5.上皮毛细血管	5. Capillary
6.角质层	6. Stratum corneum
7.颗粒层	7. Stratum granulosum
8.颗粒细胞	8. Granulosa cell
9.棘细胞层	9. Stratum spinosum
10.基底层	10. Stratum basale

图5-5 瘤胃

Fig.5-5 Rumen

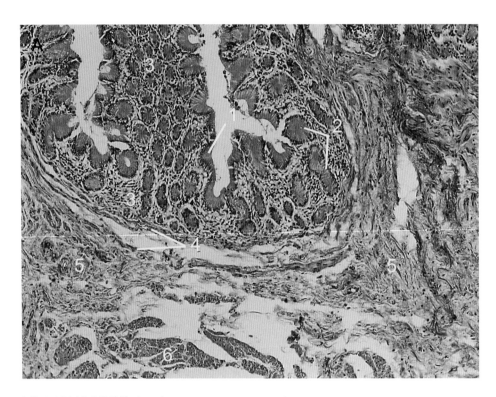

A.腺囊区固有层内充满腺体（40×）。
A. The lamina propria filled with glands (40×).

1.上皮	1. Epithelium
2.腺体	2. Gland
3.固有层	3. Lamina propria
4.黏膜肌层	4. Muscularis mucosa
5.黏膜下层	5. Submucosa
6.肌层	6. Tunica muscularis

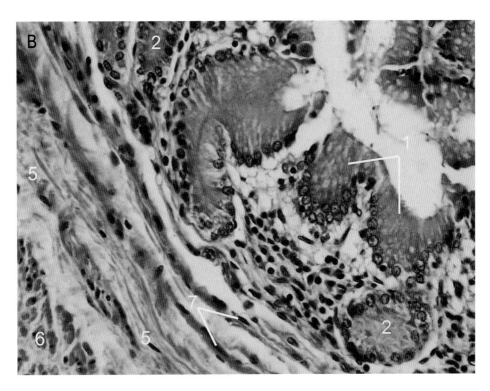

B. 腺上皮及腺囊区上皮细胞均为柱状（400×）。
B. All epithelial cells are columnar (400×).

1. 上皮	1. Epithelium
2. 腺体	2. Gland
5. 黏膜下层	5. Submucosa
6. 肌层	6. Tunica muscularis
7. 平滑肌	7. Smooth muscle

图5-6 瘤胃腺囊

Fig.5-6 Glandular sacs of rumen

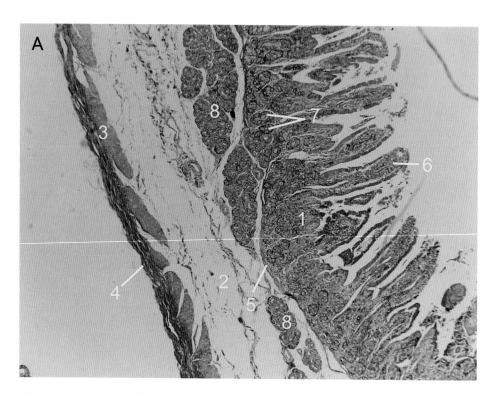

A.黏膜下层广泛分布十二指肠腺（40×）。
A. The duodenal glands are widely distributed in the submucosa (40×).

1.黏膜	1. Mucous membrane
2.黏膜下层	2. Submucosa
3.肌层	3. Tunica muscularis
4.外膜	4. External tunic
5.黏膜肌层	5. Muscularis mucosa
6.绒毛	6. Intestinal villus
7.肠腺	7. Intestinal gland
8.十二指肠腺	8. Duodenal gland

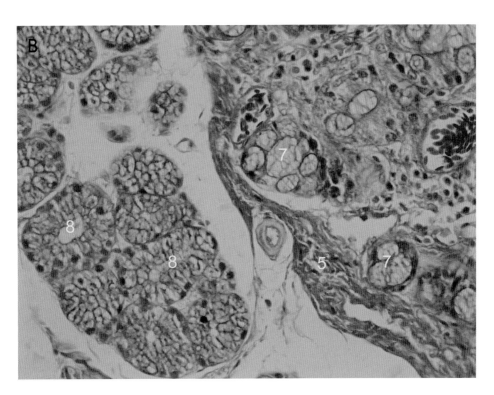

B. 十二指肠腺及小肠腺 (400×)。
B. The duodenal glands and small intestinal glands (400×).

5. 黏膜肌层	5. Muscularis mucosa
7. 肠腺	7. Intestinal gland
8. 十二指肠腺	8. Duodenal gland

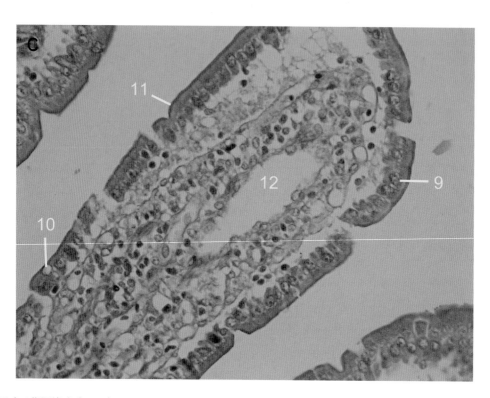

C.十二指肠绒毛（400×）。
C. The intestinal villus (400×).

9.柱状上皮	9. Simple columnar epithelium
10.杯状细胞	10. Goblet cell
11.纹状缘	11. Striated border
12.中央乳糜管	12. Central lacteal

图5-7 十二指肠
Fig.5-7 Duodenum

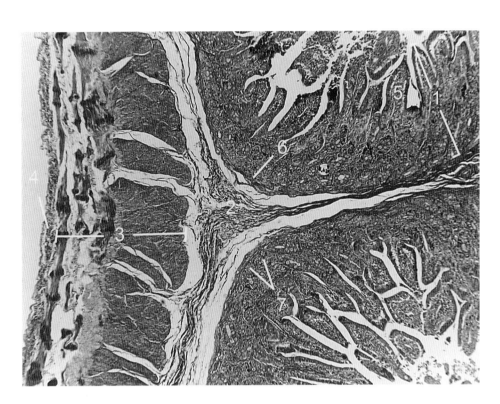

绒毛短呈叶状（40×）。
The intestinal villus are short and leaf-shaped (40×).

1.黏膜	1. Mucous membrane
2.黏膜下层	2. Submucosa
3.肌层	3. Tunica muscularis
4.外膜	4. External tunic
5.绒毛	5. Intestinal villus
6.黏膜肌层	6. Muscularis mucosa
7.肠腺	7. Intestinal gland

图5-8 空肠

Fig.5-8 Jejunum

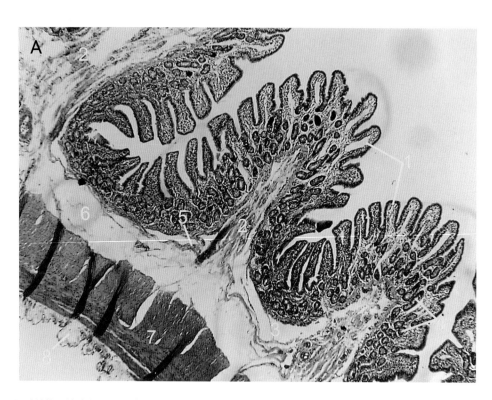

A. 肠壁结构，皱襞多而明显（40×）。
A. The ileum wall, its plica are numerous and obvious (40×).

1. 绒毛	1. Intestinal villus
2. 皱襞	2. Plica
3. 固有层	3. Lamina propria
4. 肠腺	4. Intestinal gland
5. 黏膜肌层	5. Muscularis mucosa
6. 黏膜下层	6. Submucosa
7. 肌层	7. Tunica muscularis
8. 外膜	8. External tunic

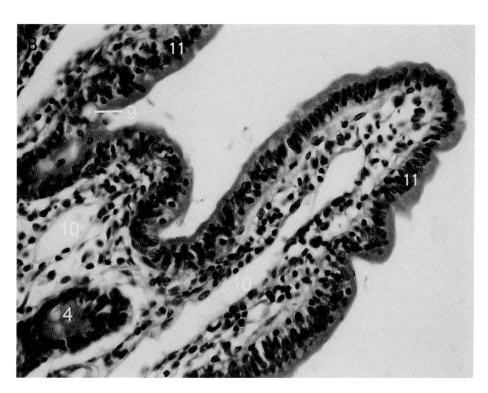

B.肠绒毛（400×）。
B. The intestinal villus (400×).

4.肠腺	4. Intestinal gland
9.杯状细胞	9. Goblet cell
10.中央乳糜管	10. Central lacteal
11.柱状细胞	11. Columnar cell

图5-9 回肠
Fig.5-9 Ileum

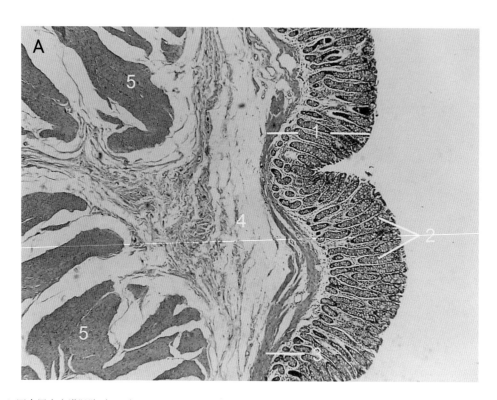

A.固有层内充满肠腺（40×）。
A. The lamina propria filled with intestinal glands (40×).

1.黏膜	1. Mucous membrane
2.肠腺	2. Intestinal gland
3.黏膜肌层	3. Muscularis mucosa
4.黏膜下层	4. Submucosa
5.内环肌层	5. Internal annular tunica muscularis

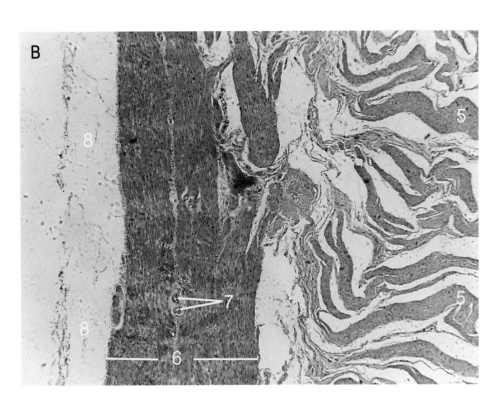

B. 肌层（40×）。
B. The tunica muscularis of cecum (40×).

5. 内环肌层	5. Internal annular tunica muscularis
6. 外纵肌层	6. External longitudinal tunica muscularis
7. 小动脉	7. Small artery
8. 外膜	8. External tunic

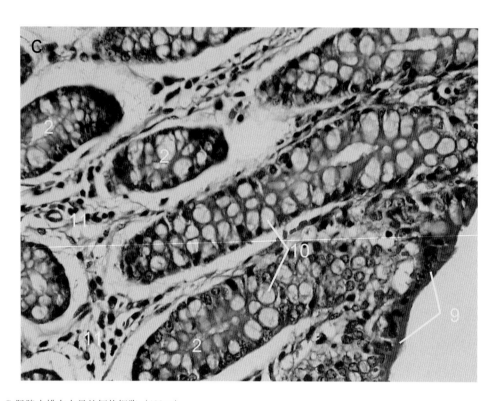

C.肠腺内排布大量的杯状细胞（400×）。
C. The goblet cells are distributed in the intestinal glands (400×).

2.肠腺	2. Intestinal gland
9.肠上皮细胞	9. Intestinal epithelial cell
10.杯状细胞	10. Goblet cell
11.固有层	11. Lamina propria

图5-10 盲肠

Fig.5-10 Cecum

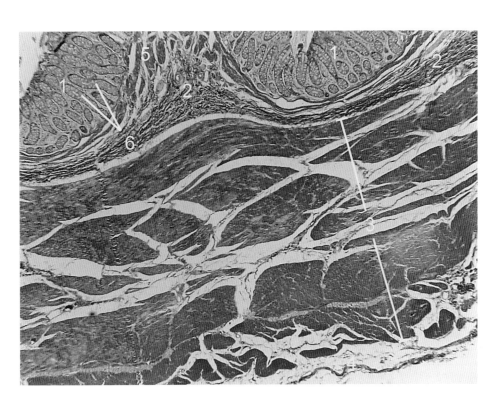

肠壁结构，肌层发达（40×）。
The colon wall, its tunica muscularis is developed (40×).

1.黏膜	1. Mucous membrane
2.黏膜下层	2. Submucosa
3.肌层	3. Tunica muscularis
4.外膜	4. External tunic
5.黏膜肌层	5. Muscularis mucosa
6.肠腺	6. Intestinal gland

图5-11 结肠
Fig.5-11 Colon

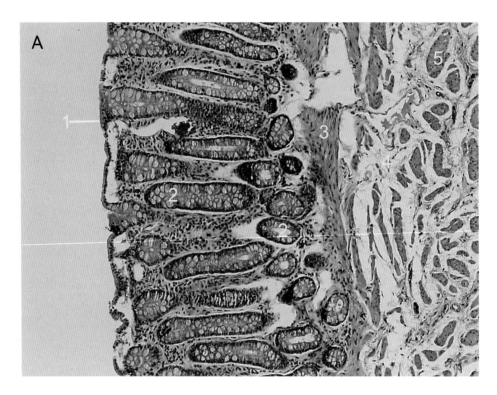

A.肠上皮表面平整、规则（100×）。
A. The epithelial surface is smooth and regular (100×).

1.上皮	1. Epithelium
2.肠腺	2. Intestinal gland
3.黏膜肌层	3. Muscularis mucosa
4.结缔组织	4. Connective tissue
5.平滑肌束	5. Smooth muscle bundles

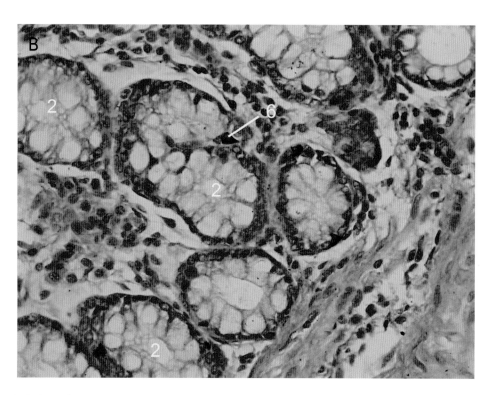

B.肠腺（400×）。
B. The intestinal glands of rectum (400×).

2.肠腺	2. Intestinal gland
6.内分泌细胞	6. Endocrine cell

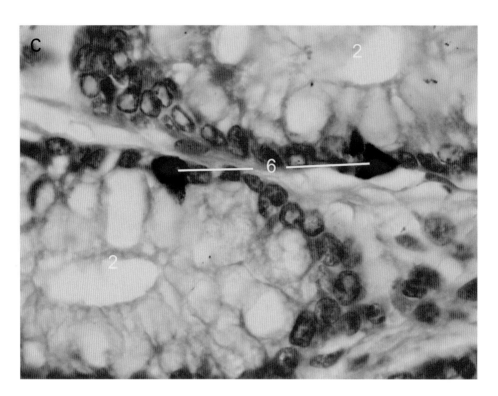

C. 内分泌细胞内有大量的黑棕色颗粒（1 000×）。
C. The endocrine cells in rectum, having plenty of black brown granules (1 000×).

2. 肠腺	2. Intestinal gland
6. 内分泌细胞	6. Endocrine cell

图5-12　直肠

Fig.5-12　Rectum

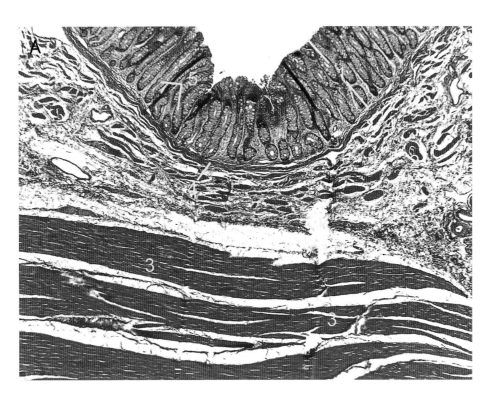

A. 黏膜下层发达（40×）。
A. The submucosa is developed (40×).

1. 黏膜	1. Mucous membrane
2. 黏膜下层	2. Submucosa
3. 肌层	3. Tunica muscularis
4. 黏膜肌层	4. Muscularis mucosa
5. 肠腺	5. Intestinal gland

B. 肌层（40×）。
B. The tunica muscularis (40×).

| 3. 肌层 | 3. Tunica muscularis |
| 6. 外膜 | 6. External tunic |

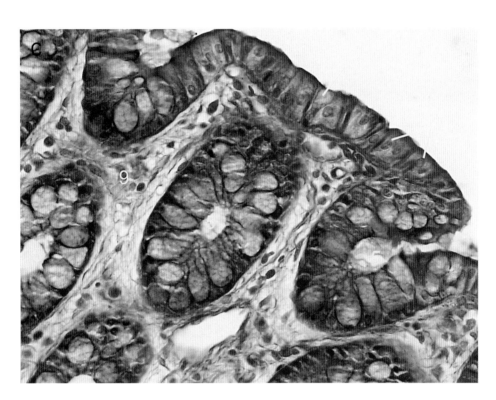

C. 上皮主要为柱状细胞（400×）。
C. The epithelium mainly consists of columnar cells (400×).

5. 肠腺	5. Intestinal gland
7. 柱状上皮	7. Columnar epithelium
8. 杯状细胞	8. Goblet cell
9. 固有层	9. Lamina propria

图 5-13　肛门
Fig.5-13　Anus

第六章 消化腺

CHAPTER 6

骆驼消化腺主要包括腮腺、下颌腺、肝脏与胰腺等独立壁外腺体，通过外分泌导管与消化管相接。消化腺分泌的消化液主要对食物进行化学性消化。此外，消化腺还具有其他方面的功能，如免疫、解毒和内分泌等。

CHAPTER 6

Digestive Glands

The digestive glands of camel consist of several components, namely, salivary gland, liver and pancreas, which have excretory ducts that open into the digestive tract. Digestive fluid secreted by digestive glands is mainly used for chemical digestion of food. In addition, digestive glands have other functions such as immune, detoxification and endocrine.

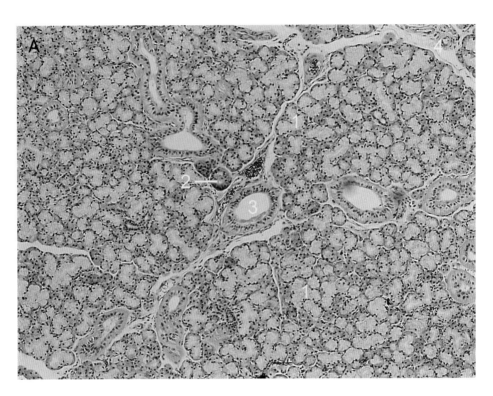

A.实质内可见大量的黏液性腺泡（40×）。
A. There are plenty of mucosa alveoli in the parenchyma （40×）.

1.黏液性腺泡	1. Mucosa alveoli
2.浆液性腺泡	2. Serous alveoli
3.导管	3. Duct
4.小叶间结缔组织	4. Interlobular connective tissue

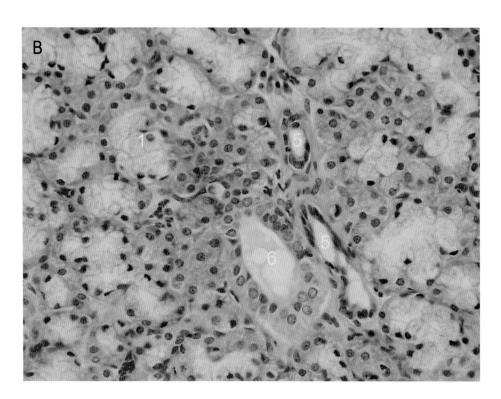

B.腺泡间导管清晰可见（40×）。
B. The glandular ducts (40×).

1.黏液性腺泡	1. Mucosa alveoli
5.闰管	5. Intercalated duct
6.纹管	6. Striated duct

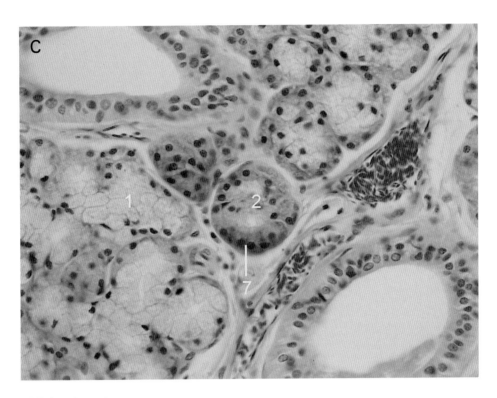

C.浆液性半月 (400×)。
C. The serous demilune (400×).

1.黏液性腺泡	1. Mucosa alveoli
2.浆液性腺泡	2. Serous alveoli
7.浆液性半月	7. Serous demilune

图6-1 下颌腺

Fig.6-1 Submandibular gland

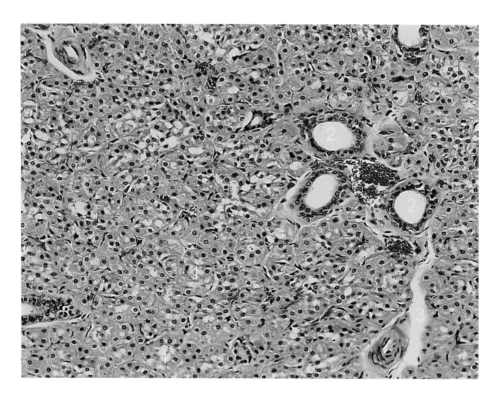

实质内可见大量的黏液性腺泡（200×）。
There are many mucosa alveoli in the parenchyma (200×).

1. 分泌单位	1. Secretory unit
2. 纹管	2. Striated duct
3. 小叶间结缔组织	3. Interlobular connective tissue

图6-2 腮腺

Fig.6-2 Parotid

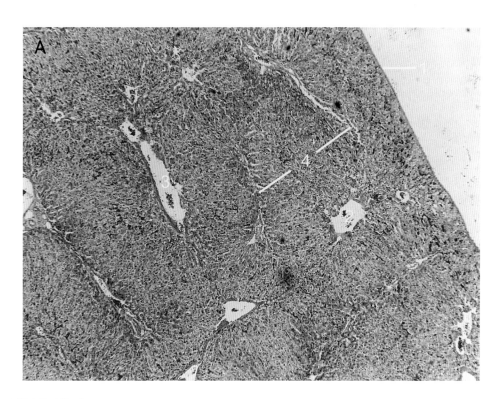

A.肝小叶（40×）。
A. The hepatic lobule (40×).

1. 被膜	1. Capsule
2. 中央静脉（横断面）	2. Central vein (transverse section)
3. 中央静脉（纵断面）	3. Central vein (vertical section)
4. 肝小叶	4. Hepatic lobule

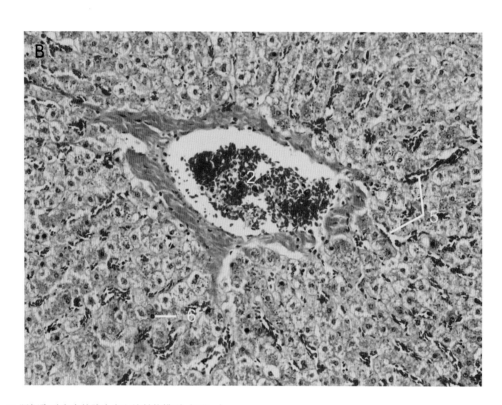

B.肝细胞以中央静脉为中心放射状排列（200×）。
B. Hepatocytes are arranged radially around the central vein（200×）.

2.中央静脉（横断面）	2. Central vein（transverse section）
5.肝血窦	5. Hepatic sinusoid
6.肝细胞	6. Hepatocyte

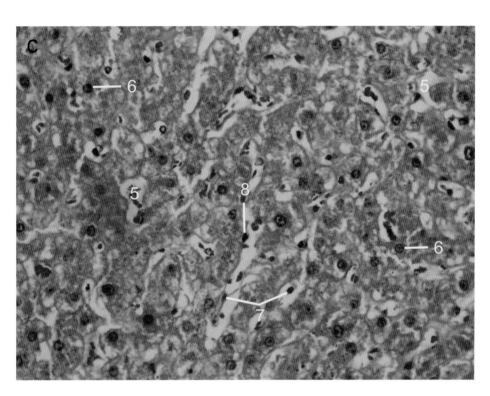

C.肝小叶结构（400×）。
C. The component of hepatic lobule (400×).

5.肝血窦	5. Hepatic sinusoid
6.肝细胞	6. Hepatocyte
7.枯否氏细胞	7. Kupffer cell
8.内皮细胞	8. Endothelial cell

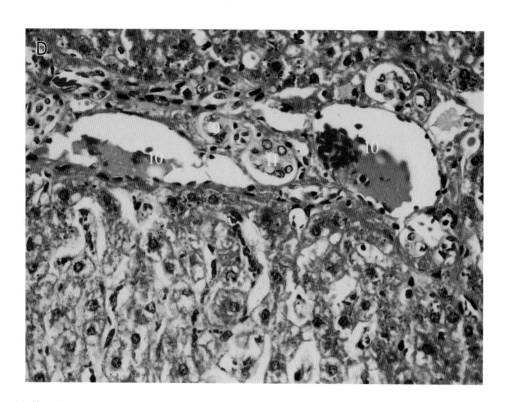

D. 门管区（400×）。
D. The portal area (400×).

9. 小叶间动脉	9. Interlobular artery
10. 小叶间静脉	10. Interlobular vein
11. 小叶间胆管	11. Bile duct

图6-3 肝脏
Fig.6-3 Liver

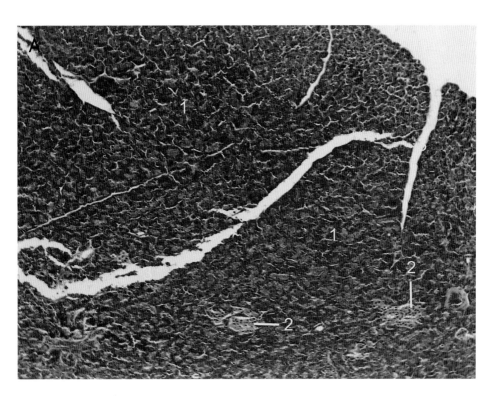

A. 外分泌部和胰岛（100×）。
A. The exocrine portion and pancreatic islet (100×).

1. 外分泌部	1. Exocrine portion
2. 胰岛	2. Pancreatic islet

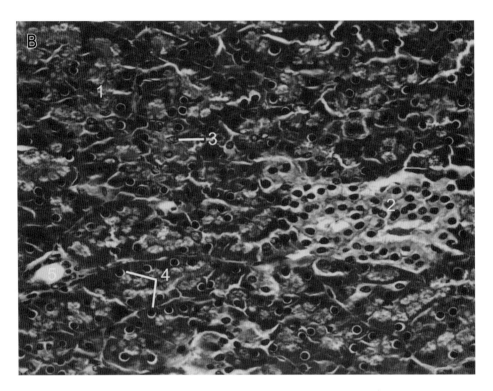

B.腺泡和闰管，腺泡细胞顶部嗜酸性，基部嗜碱性（400×）。
B. The acinus and intercalated duct, acinus cells are acidophilic at the top and basophilic at the base （400×）.

1.外分泌部	1. Exocrine portion
2.胰岛	2. Pancreatic islet
3.泡心细胞	3. Centroacinar cell
4.腺泡细胞	4. Acinus cell
5.闰管	5. Intercalated duct

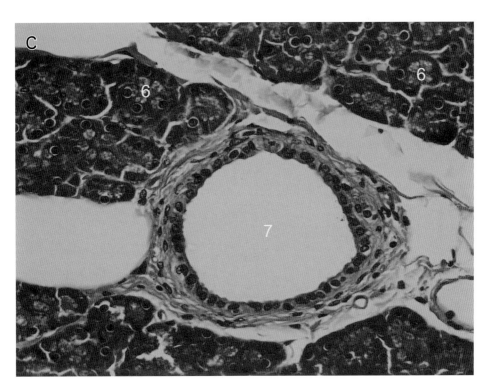

C. 小叶内导管（400×）。
C. The intralobular duct (400×).

| 6. 腺泡 | 6. Acinus |
| 7. 小叶内导管 | 7. Intralobular duct |

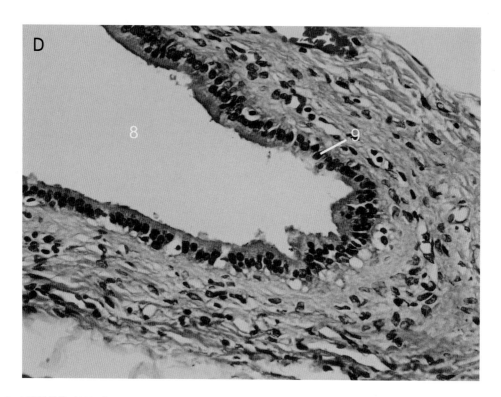

D. 小叶间导管（400×）。
D. The interlobular duct (400×).

| 8. 小叶间导管 | 8. Interlobular duct |
| 9. 单层柱状上皮 | 9. Simple columnar epithelium |

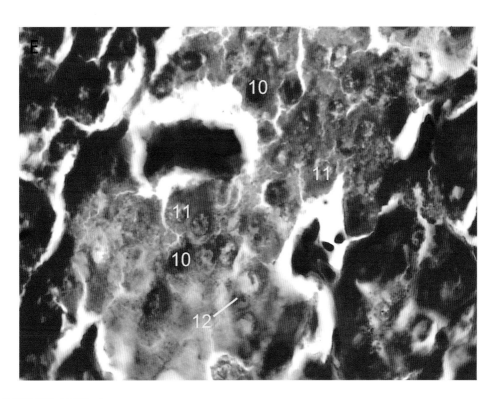

E. 胰岛细胞（1 000×）。
E. The cells in the pancreatic islet (1 000×).

10. A 细胞	10. A cell
11. B 细胞	11. B cell
12. D 细胞	12. D cell

图 6-4　胰腺

Fig.6-4 Pancreas

第七章 呼吸系统

CHAPTER 7

骆驼呼吸系统由肺和连接肺组织与外部环境的系列管道组成。该系统具有传输气体与呼吸功能的两部分结构，共同执行气体交换功能。在本章中，只描述鼻孔、支气管与肺的组织学结构。

CHAPTER 7

Respiratory System

The respiratory system of camel is composed of the lung and a series of tracts linking the pulmonary tissue with the external environment. This system is customarily divided into the conducting and respiratory portions to perform gas exchange. In the present chapter, histology of nostrils, bronchus and lungs only will be described.

A.管壁内可见大量的腺体及骨骼肌纤维（40×）。
A. The nostril wall, with plenty of glands and skeletal muscle fibers (40×).

1.复层扁平上皮	1. Stratified squamous epithelium
2.固有层	2. Lamina propria
3.骨骼肌	3. Skeletal muscle
4.腺体	4. Gland
5.导管	5. Duct

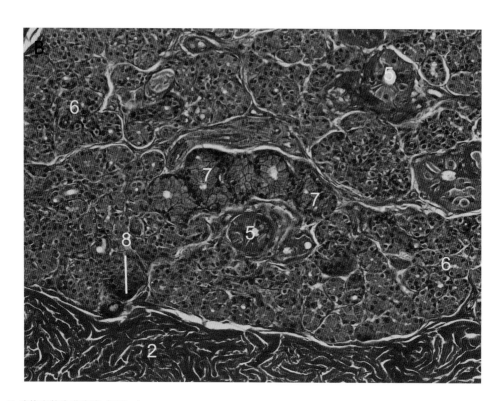

B.腺体多数为黏液腺（200×）。
B. Mainly mucous alveoli (200×).

2.固有层	2. Lamina propria
5.导管	5. Duct
6.浆液性腺泡	6. Serous alveoli
7.黏液性腺泡	7. Mucous alveoli
8.浆液性半月	8. Serous demilune

图 7-1　鼻孔

Fig.7-1 Nostril

A.管壁结构（40×）。
A. The bronchial wall (40×).

1.假复层纤毛柱状上皮	1. Pseudostratified ciliated columnar epithelium
2.固有层	2. Lamina propria
3.平滑肌束	3. Smooth muscle bundle
4.软骨片	4. Cartilage slice
5.外膜	5. Tunica adventitia

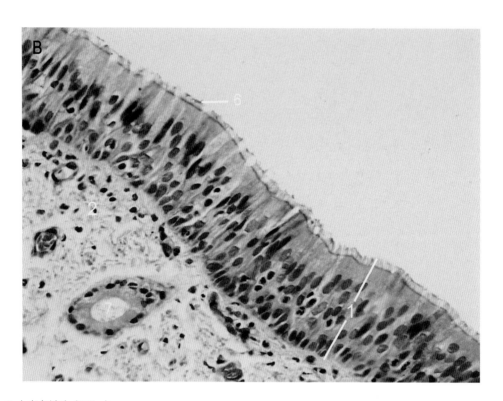

B. 上皮有纤毛（400×）。
B. The ciliated epithelium (400×).

1. 假复层纤毛柱状上皮	1. Pseudostratified ciliated columnar epithelium
2. 固有层	2. Lamina propria
6. 纤毛	6. Cilium
7. 气管腺	7. Tracheal gland

图7-2 支气管
Fig.7-2 Bronchus

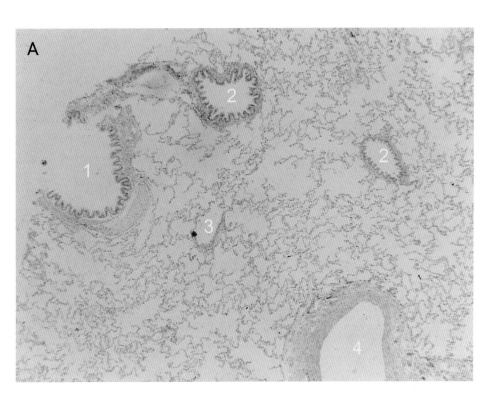

A.肺导气部（40×）。
A. The conducting portion (40×).

1.小支气管	1. Bronchium
2.细支气管	2. Bronchiole
3.终末细支气管	3. Terminal bronchiole
4.血管	4. Blood vessel

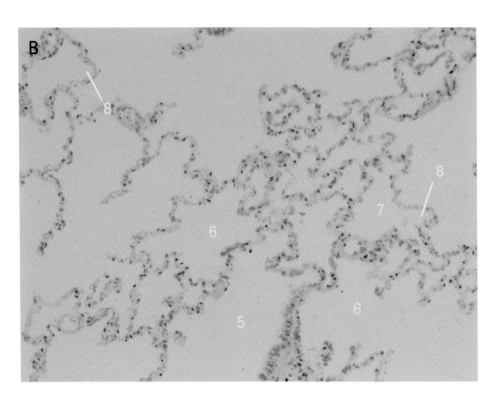

B.肺呼吸部（200×）。
B. The respiratory portion (200×).

5.呼吸性细支气管	5. Respiratory bronchiole
6.肺泡管	6. Alveolar duct
7.肺泡囊	7. Alveolar sac
8.肺泡	8. Pulmonary alveoli

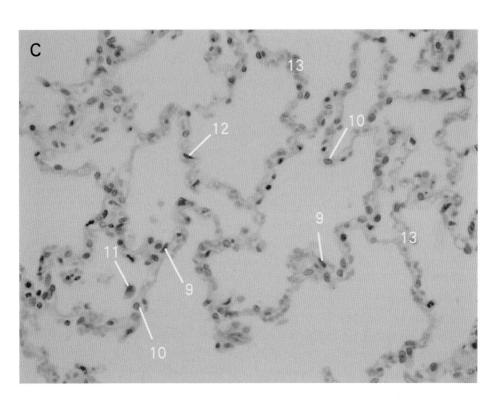

C.肺泡结构（400×）。
C. The detailed alveolar structure (400×).

9. Ⅰ型肺泡细胞	9. Type Ⅰ (alveolar) cell
10. Ⅱ型肺泡细胞	10. Type Ⅱ (alveolar) cell
11. 尘细胞	11. Dust cell
12. 内皮细胞	12. Endothelial cell
13. 肺泡隔	13. Alveolar septum

图7-3　肺
Fig.7-3　Lung

第八章 泌尿系统

CHAPTER 8

骆驼泌尿系统包括两个肾、两条输尿管、一个膀胱与一条尿道，其主要功能是生成、储存与排出尿液。肾脏是泌尿系统的重要器官，生成尿液；输尿管是两条纤维肌性管道，将尿液引导至膀胱储存；尿道连接膀胱与外界。

CHAPTER 8

Urinary System

The urinary system of camel includes two kidneys, two ureters, a bladder, and a urethra. Its main function is to produce, store and excrete the urine. The kidneys are the important organ of the urinary system, which elaborate a fluid product called urine; the ureters, two fibromuscular tubes, conduct the urine to a single urinary bladder where the fluid accumulates for periodic evacuation via the single urethra that connects the bladder to the exterior.

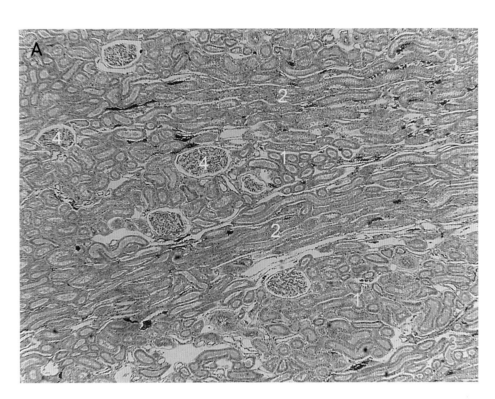

A. 肾皮质及部分髓质，髓放线与皮质迷路相互交替分布（40×）。
A. The cortex and medulla, the alternating distribution of medullary ray and cortical labyrinth (40×).

1. 皮质迷路	1. Cortical labyrinth
2. 髓放线	2. Medullary ray
3. 髓质	3. Medulla
4. 肾小体	4. Renal corpuscle

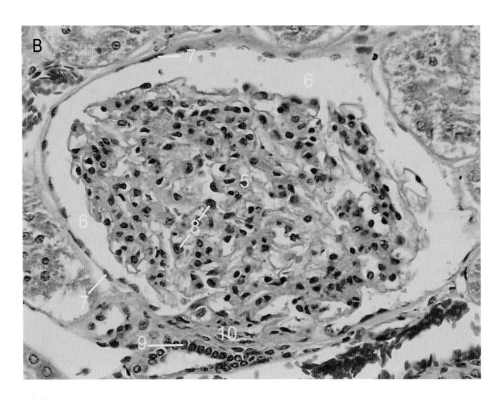

B.肾小体（400×）。
B. The renal corpuscle (400×).

5.血管球	5. Glomerulus
6.肾小囊腔	6. Capsular space
7.肾小囊壁层	7. Capsular epithelium
8.足细胞	8. Podocyte
9.致密斑	9. Macula densa
10.球外系膜细胞	10. Extraglomerular mesangial cell

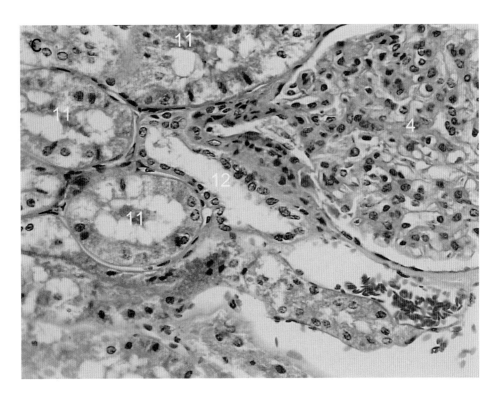

C.近曲小管管腔不规则，管壁细胞胞质嗜酸性（400×）。
C. The proximal convoluted tubule with irregular lumen and the acidophilic cells (400×).

4.肾小体	4. Renal corpuscle
11.近曲小管	11. Proximal convoluted tubule
12.远曲小管	12. Distal convoluted tubule

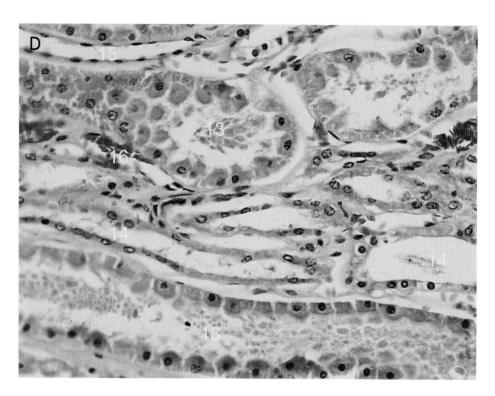

D.髓质，泌尿小管不同部位结构（400×）。
D. The medulla, the uriniferous tubule (400×).

13.近直小管	13. Proximal straight tubule
14.远直小管	14. Distal straight tubule
15.细段	15. Thin segment
16.毛细血管	16. Capillary

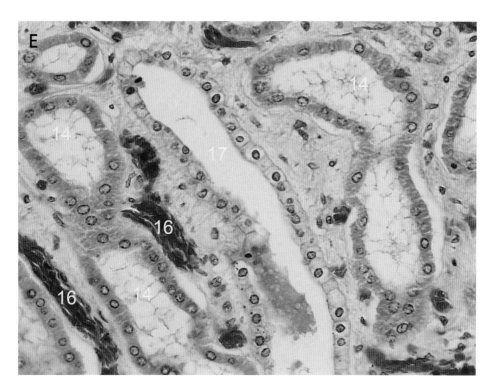

E.髓质，集合小管细胞界限明显（400×）。
E. The medulla, there is a clear line of cells in the collecting tubules （400×）.

14.远直小管	14. Distal straight tubule
16.毛细血管	16. Capillary
17.集合小管	17. Collecting tubule

图8-1 肾

Fig.8-1 Kidney

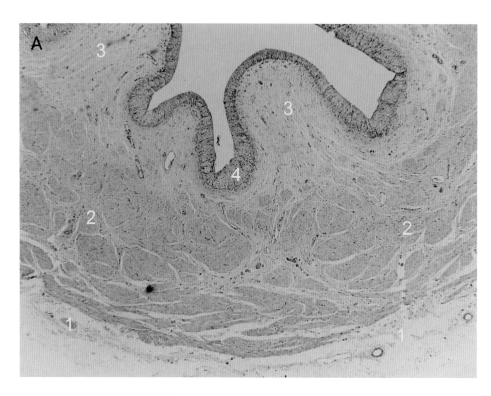

A.肌层由内纵外环平滑肌构成（40×）。
A. The tunica muscularis consists of internal longitudinal and external annular smooth muscle layers (40×).

1. 外膜	1. Tunica adventitia
2. 肌层	2. Tunica muscularis
3. 固有层	3. Lamia propria
4. 变移上皮	4. Transitional epithelium

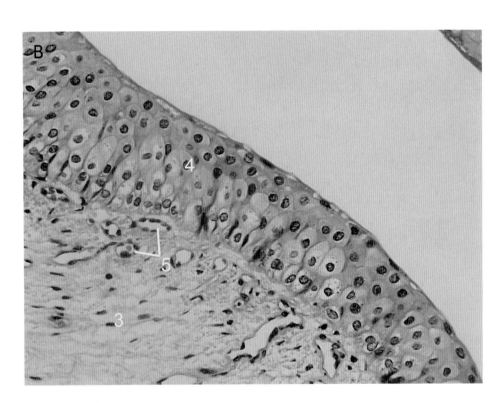

B. 变移上皮（400×）。
B. The transitional epithelium (400×).

3. 固有层	3. Lamia propria
4. 变移上皮	4. Transitional epithelium
5. 毛细血管	5. Capillary

图 8-2　输尿管
Fig.8-2　Ureter

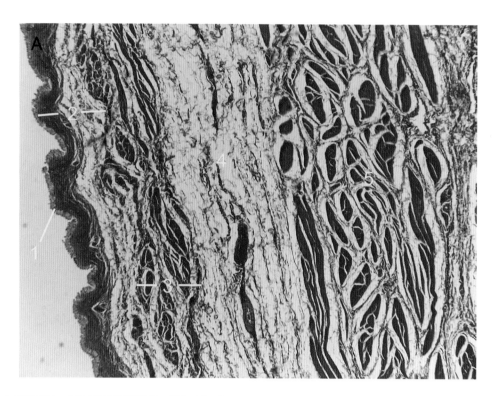

A.肌层很厚，此图仅显示了肌层的一部分（40×）。
A. Part of the tunica muscularis (40×).

1.黏膜上皮	1. Epithelium mucosa
2.固有层	2. Lamia propria
3.黏膜肌层	3. Muscularis mucosa
4.黏膜下层	4. Tunica submucosa
5.肌层	5. Tunica muscularis

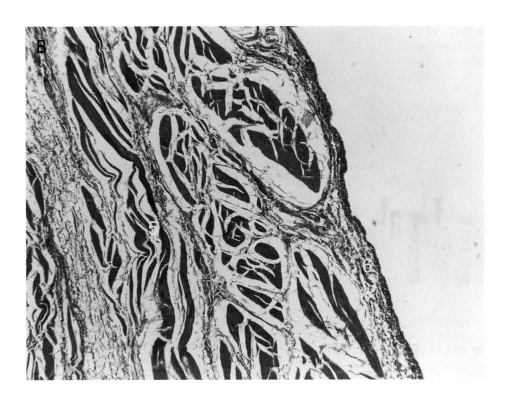

B.部分肌层和浆膜（40×）。
B. The tunic serosa and part of the tunica muscularis (40×).

5.肌层	5. Tunica muscularis
6.浆膜	6. Tunica serosa

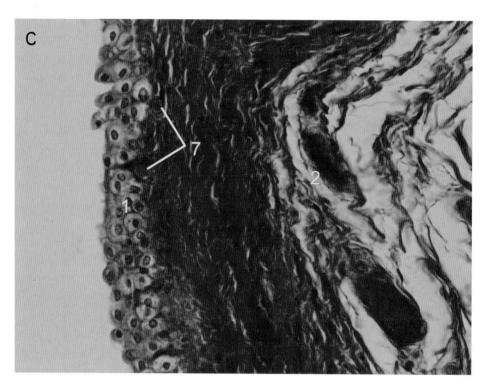

C.黏膜变移上皮下方分布大量的毛细血管（400×）。
C. Plenty of capillaries are distributed beneath the transitional epithelium （400×）.

1.黏膜上皮	1. Epithelium mucosa
2.固有层	2. Lamia propria
7.毛细血管	7. Capillary

图 8-3　膀胱

Fig.8-3 Urinary bladder

第九章 雄性生殖系统

CHAPTER 9

雄性骆驼的生殖系统包括睾丸、附睾、输精管、副性腺、尿道与阴茎。睾丸是位于阴囊的成对器官，具有产生精子和分泌雄性激素的双重功能。附睾附着在睾丸的一侧，对精子的发育和成熟起作用。副性腺，如前列腺和尿道球腺，促进精液形成。

CHAPTER 9

Male Reproductive System

The male reproductive system of camel includes testes, the epididymis, the ductus deferens, the accessary glands, the urethral and finally the penis. The testes are paired organs that line in the scrotal sac, have a dual function for production of spermatozoa, and secretion of male sex hormones, principally testosterone. The epididymis attaches on the one side of testes to play the role for the development and maturation of spermatozoa. The accessory glands, such as prostate and bulbourethral glands, contribute to the formation of seminal plasma.

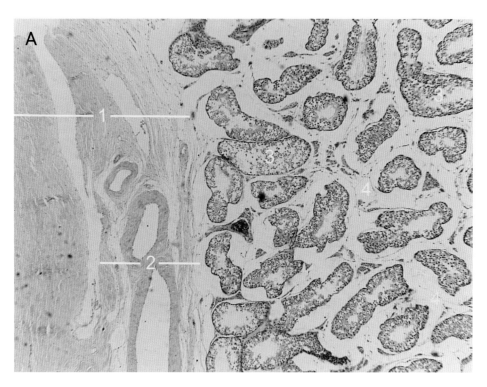

A.白膜及曲精小管,白膜内含有丰富的血管（40×）。
A. The tunica albuginea and seminiferous tubules, the tunica albuginea containing plenty of blood vessels（40×）.

1.白膜	1. Tunica albuginea
2.白膜血管	2. Blood vessels in tunica albuginea
3.曲精小管	3. Seminiferous tubule
4.睾丸间质	4. Interstitium testis

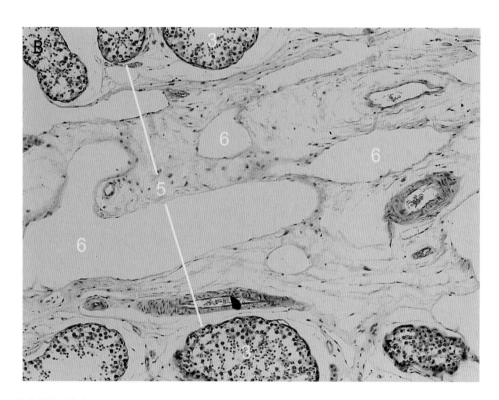

B.睾丸纵隔及睾丸网（100×）。
B. The mediastinum testis and rete testis (100×).

3.曲精小管	3. Seminiferous tubule
5.睾丸纵隔	5. Mediastinum testis
6.睾丸网	6. Rete testis

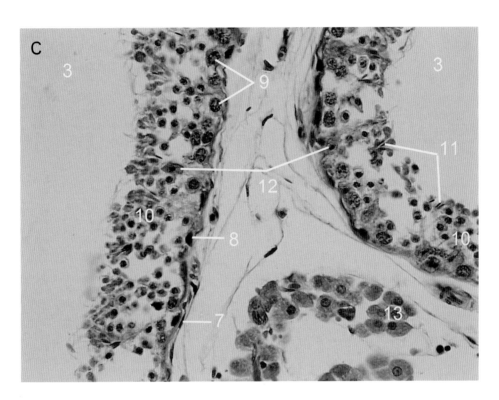

C.曲精小管，间质中富含间质细胞（400×）。
C. The seminiferous tubules, interstitial cells are densely distributed in the interstitium （400×）.

3.曲精小管	3. Seminiferous tubule
7.肌样细胞	7. Myoid cell
8.精原细胞	8. Spermatogonia
9.初级精母细胞	9. Primary spermatocyte
10.精子细胞	10. Spermatid
11.精子	11. Spermatozoon
12.支持细胞	12. Sustentacular cell
13.间质细胞	13. Interstitial cell

图9-1　睾丸
Fig.9-1　Testis

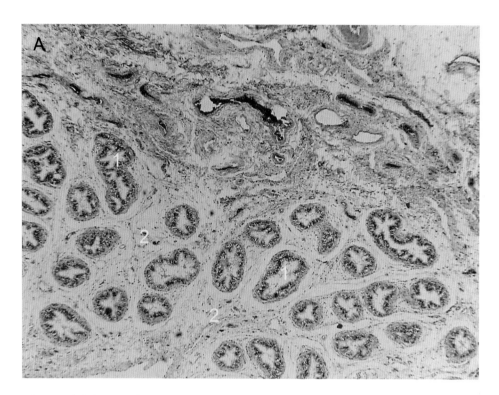

A.附睾头部，睾丸输出小管（40×）。
A. The epididymal head, the efferent ducts (40×).

1.睾丸输出小管	1. Efferent duct
2.疏松结缔组织	2. Loose connective tissue

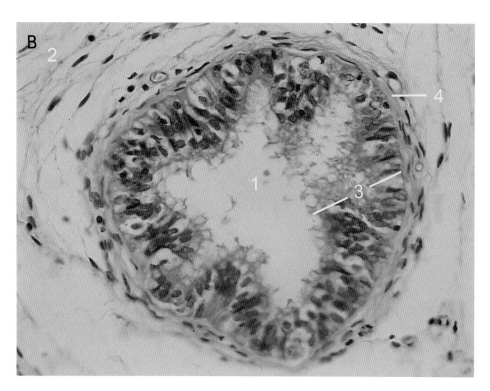

B. 睾丸输出小管管壁高处的假复层纤毛柱状上皮（400×）。
B. The pseudostratified ciliated columnar epithelium located in the protrusion of the efferent ducts (400×).

1. 睾丸输出小管	1. Efferent duct
2. 疏松结缔组织	2. Loose connective tissue
3. 假复层上皮	3. Pseudostratified epithelium
4. 平滑肌	4. Smooth muscle

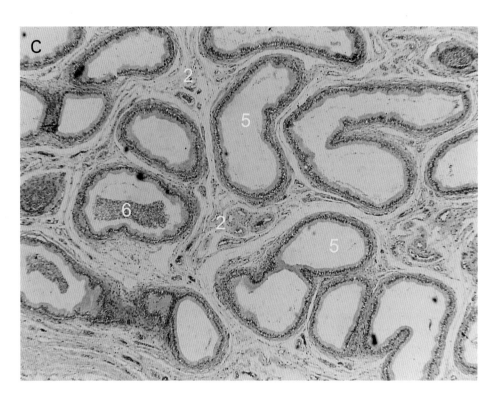

C.附睾体部，附睾管（40×）。
C. The epididymal body, the epididymal duct (40×).

2.疏松结缔组织	2. Loose connective tissue
5.附睾管	5. Epididymal duct
6.精子	6. Spermatozoon

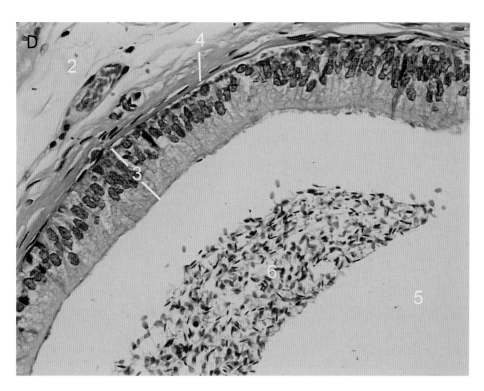

D.附睾管管壁假复层柱状上皮（400×）。
D. The pseudostratified ciliated columnar epithelium （400×）.

2.疏松结缔组织	2. Loose connective tissue
3.假复层上皮	3. Pseudostratified epithelium
4.平滑肌	4. Smooth muscle
5.附睾管	5. Epididymal duct
6.精子	6. Spermatozoon

图 9-2　附睾

Fig. 9-2 Epididymis

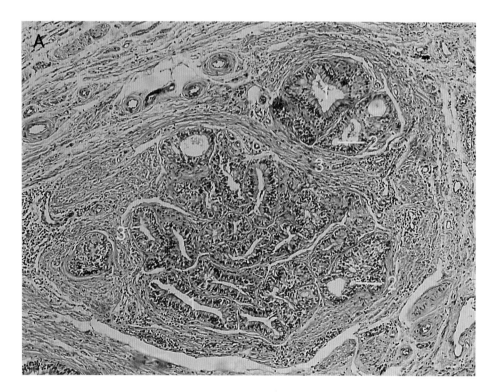

A. 腺泡由小叶间隔分开，小叶间隔内富含平滑肌（100×）。
A. The glands are separated by the interlobular septum, which contains smooth muscles (100×).

1. 腺体	1. Gland
2. 导管	2. Duct
3. 小叶间隔	3. Interlobular septum

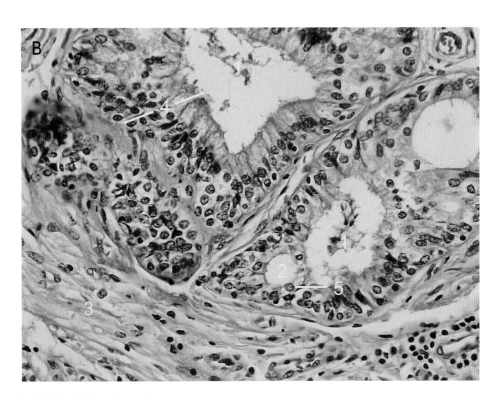

B.腺泡上皮和导管上皮（100×）。
B. The glandular and duct epithelium （100×）.

1. 腺体	1. Gland
2. 导管	2. Duct
3. 小叶间隔	3. Interlobular septum
4. 假复层柱状上皮	4. Pseudostratified columnar epithelium
5. 立方上皮	5. Cuboidal epithelium

图9-3 精囊腺

Fig.9-3 Vesicular gland

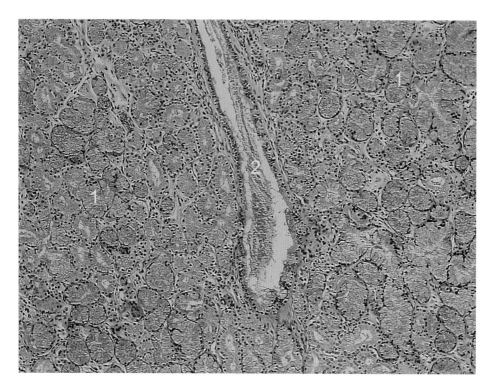

由黏液性腺泡组成（100×）。
Containing the mucous alveoli（100×）.

1. 黏液性腺泡	1. Mucous alveoli
2. 导管	2. Duct

图9-4　尿道球腺

Fig.9-4 Bulbourethral gland

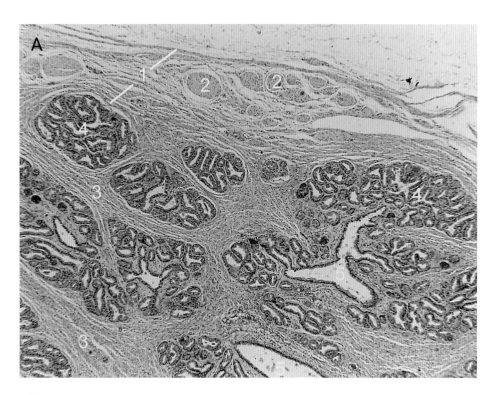

A.被膜发达，小梁将腺体分成许多小叶（40×）。
A. The capsule is developed, and the gland is separated into many lobules by the trabecula (40×).

1. 被膜	1. Capsule
2. 神经	2. Nerve
3. 小梁	3. Trabecula
4. 腺体	4. Gland
5. 导管	5. Duct

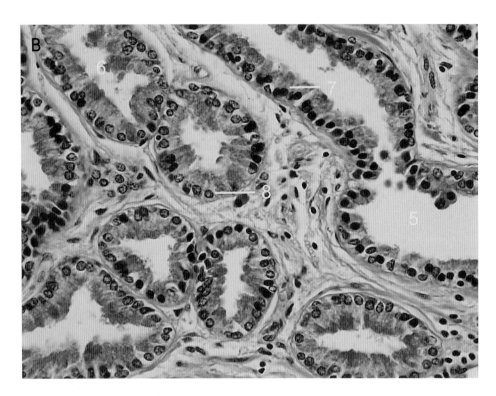

B. 浆液腺（400×）。
B. Prostate is the serous gland（400×）.

5. 导管	5. Duct
6. 分泌物	6. Secretion
7. 导管上皮细胞	7. Duct epithelial cell
8. 腺上皮细胞	8. Epithelium

图 9-5 前列腺

Fig.9-5 Prostate gland

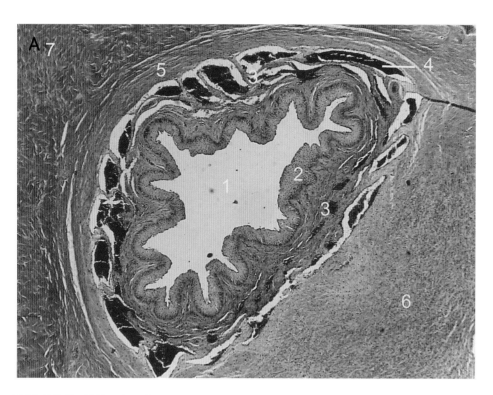

A.尿道海绵体部，横断面（40×）。
A. The cross-section views of corpus cavernosum penis (40×).

1.尿道	1. Urethra
2.尿道上皮	2. Urethra epithelium
3.尿道海绵体	3. Corpus cavernosum urethrae
4.海绵体间隙	4. Cavernous space
5.白膜	5. Tunica albuginea
6.致密结缔组织	6. Dense connective
7.阴茎海绵体	7. Corpus cavernosum penis

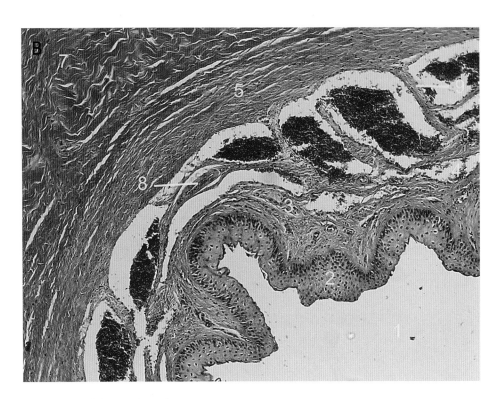

B.海绵体间隙的分布很明显（100×）。
B.The cavernous space is clear (100×).

1.尿道	1. Urethra
2.尿道上皮	2. Urethra epithelium
3.尿道海绵体	3. Corpus cavernosum urethrae
4.海绵体间隙	4. Cavernous space
5.白膜	5. Tunica albuginea
7.阴茎海绵体	7. Corpus cavernosum penis
8.血管	8. Blood vessel
9.结缔组织小梁	9. Connective tissue (trabecula of)

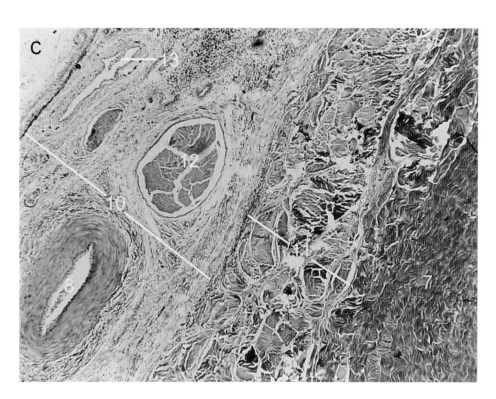

C. 外膜及阴茎缩肌，外膜富含血管及神经（40×）。
C. The tunica externa and retractor penis muscle, the capsule is rich in blood vessels and nerves (40×).

7. 阴茎海绵体	7. Corpus cavernosum penis
8. 血管	8. Blood vessel
10. 外膜	10. Tunica externa
11. 阴茎缩肌	11. Retractor penis muscle
12. 神经	12. Nerve
13. 静脉瓣	13. Vein valve

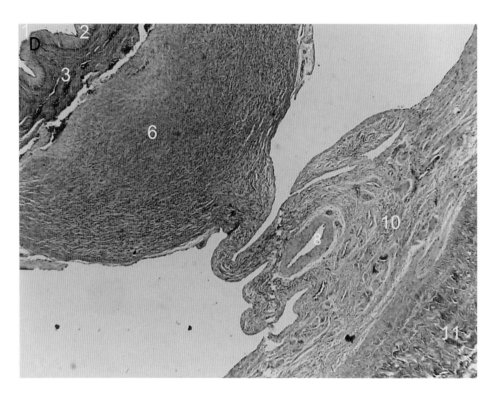

D.外膜发达，富含血管及神经（40×）。
D. The developed capsule is rich in the blood vessels and nerves (40×).

1.尿道	1. Urethra
2.尿道上皮	2. Urethra epithelium
3.尿道海绵体	3. Corpus cavernosum urethrae
6.致密结缔组织	6. Dense connective
8.血管	8. Blood vessel
10.外膜	10. Tunica adventitia
11.阴茎缩肌	11. Retractor penis muscle

图 9-6　阴茎

Fig.9-6 Penis

第十章 雌性生殖系统

CHAPTER 10

雌性骆驼的生殖系统包括卵巢、输卵管、子宫、阴道和外生殖器。卵巢产生的卵子在输卵管中受精后，输送到子宫。子宫是孕育胎儿的器官。

CHAPTER 10

Female Reproductive System

The female reproductive system of camel consists of two ovaries, two oviducts, the uterus, the vagina, and the external genitalia. The oocytes produced in the ovaries are fertilized in the oviduct and then carried to the uterus, which breeds the fetus.

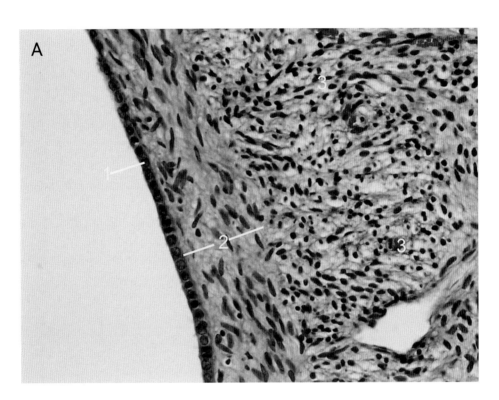

A. 被膜及皮质，生殖上皮为立方上皮（200×）。
A. The capsule and cortex, the germinal epithelium belongs to simple cuboidal epithelium （200×）.

1. 生殖上皮	1. Germinal epithelium
2. 白膜	2. Tunica albuginea
3. 皮质	3. Cortex

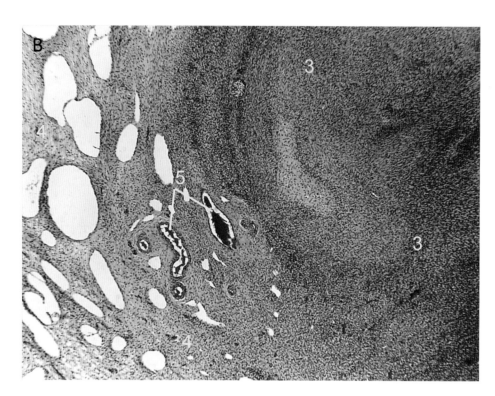

B. 卵巢皮质及髓质的一部分。髓质内含有大量的血管（400×）。
B. The cortex and medulla, many blood vessels locate in the medulla （400×）.

3. 皮质	3. Cortex
4. 髓质	4. Medulla
5. 血管	5. Blood vessel

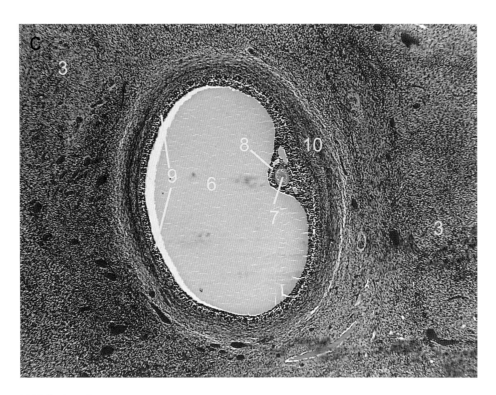

C. 次级卵泡（40×）。
C. The tertiary follicle (40×).

3. 皮质	3. Cortex
6. 卵泡腔	6. Follicular antrum
7. 卵母细胞	7. Oocyte
8. 卵丘细胞层	8. Cumulus layer
9. 颗粒细胞	9. Granular cell
10. 卵泡膜	10. Theca folliculi

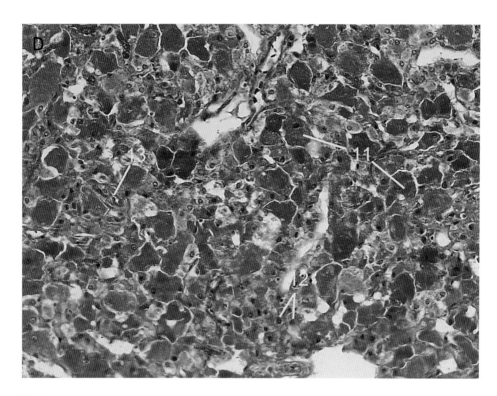

D. 黄体（200×）。
D. The corpora luteum (200×).

11.粒性黄体细胞	11. Granular lutein cell
12.膜性黄体细胞	12. Theca lutein cell
13.小动脉	13. Small artery

图10-1　卵巢
Fig.10-1　Ovary

A.肌层和外膜都较厚（40×）。
A. The muscular layer and tunica adventitia are thicker than other layers (40×).

1.黏膜	1. Mucosa
2.肌层	2. Muscular layer
3.外膜	3. Tunica adventitia
4.黏膜皱襞	4. Mucosa plica

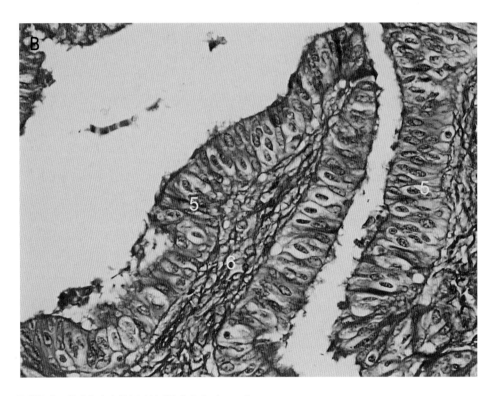

B. 黏膜皱襞，黏膜上皮为假复层纤毛柱状上皮（400×）。
B. The mucosa plica, the epithelium is the pseudostratified ciliated columnar epithelium （400×）.

5.黏膜上皮	5. Epithelium mucosa
6.结缔组织	6. Connective tissue

图10-2 输卵管

Fig.10-2 Oviduct uterine tube

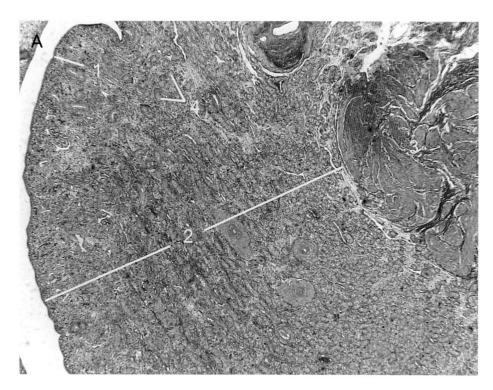

A. 黏膜固有层内分布子宫腺体（40×）。
A. The uterine glands distributed in lamia propria (40×).

1. 上皮	1. Epithelium
2. 固有层	2. Lamia propria
3. 肌层	3. Muscular layer
4. 子宫腺	4. Uterine gland

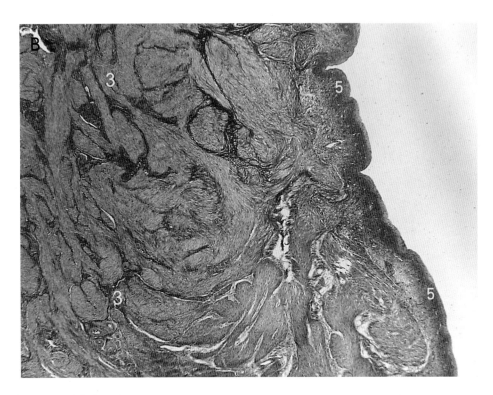

B. 部分肌层和外膜（40×）。
B. Part of muscular layer and the tunica serosa (40×).

3. 肌层　　　　　　　　　　　3. Muscular layer
5. 外膜　　　　　　　　　　　5. Tunica serosa

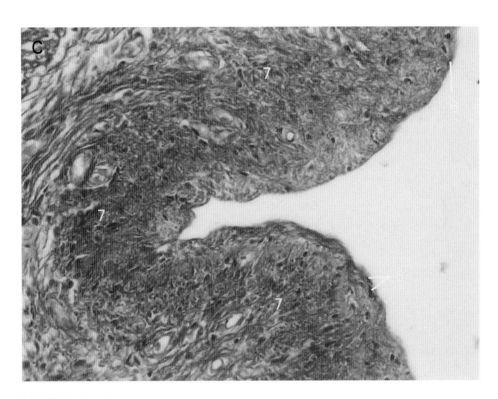

C.外膜为浆膜（400×）。
C. The capsule is tunica serosa（400×）.

6.单层扁平上皮　　　　　　　　　　6. Simple squamous epithelium

7.结缔组织　　　　　　　　　　　　7. Connective tissue

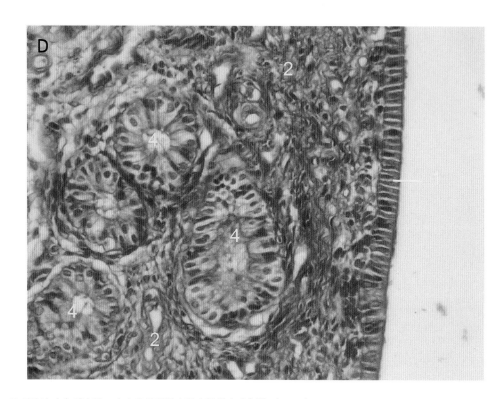

D. 子宫上皮与子宫腺，上皮为排列整齐的高柱状上皮细胞（400×）。
D. The epithelium and uterine glands, the high columnar epithelial cells are arranged neatly （400×）.

1. 上皮	1. Epithelium
2. 固有层	2. Lamia propria
4. 子宫腺	4. Uterine gland

图 10-3　子宫

Fig.10-3 Uterus

第十一章 被皮系统与骨骼肌

CHAPTER 11

骆驼被皮系统由皮肤及其附属结构组成。皮肤覆盖体表，具有保护、感觉、吸收、排泄和调节体温等功能，并参与机体免疫反应与物质代谢。骨骼肌由成束的骨骼肌细胞组成，骨骼肌细胞核位于细胞下方，具有支配躯体运动的功能。

CHAPTER 11

Integumentary System and Skeletal Muscles

The integumentary system of camel consists of the skin and epidermal derivations. The skin covers the outer surface of the body, possesses many functions, such as protection, sensory, absorption, excretion, and thermoregulation, and participates in immune responses and metabolism. The skeletal muscles consists of bundles of long and cylindrical fibers which contain a lot of peripherally placed nuclei, which function is to regulate the body movement.

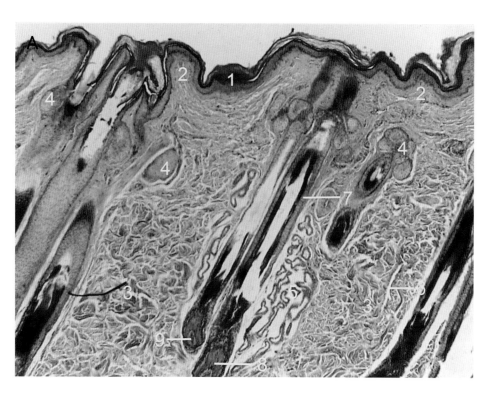

A. 皮肤及其衍生物（40×）。
A. The skin and epidermal derivatives (40×).

1. 表皮	1. Epidermis
2. 真皮乳头层	2. Dermal papillary
3. 真皮网状层	3. Dermal reticular layer
4. 皮脂腺	4. Sebaceous gland
5. 汗腺	5. Sweat gland
6. 血管	6. Blood vessel
7. 毛囊	7. Hair follicle
8. 毛球	8. Bulbus pili
9. 毛乳头	9. Papilla pili

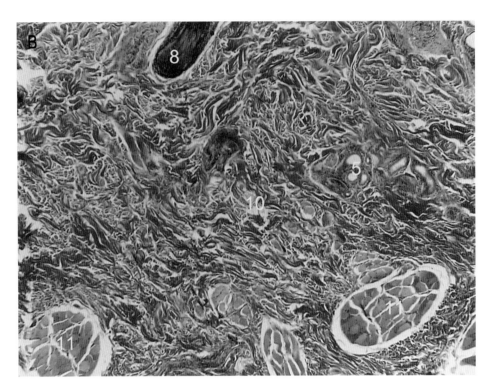

B. 皮下组织，皮下组织内可见骨骼肌纤维束（40×）。
B. The hypodermis, having skeletal muscle fiber bundles （40×）.

5. 汗腺	5. Sweat gland
8. 毛球	8. Bulbus pili
10. 皮下组织	10. Hypodermis
11. 皮肌	11. Skeletal muscle in hypodermis

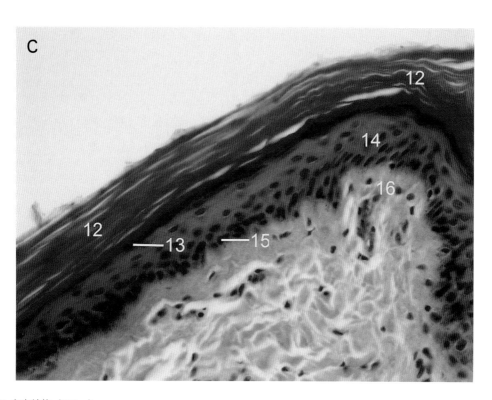

C. 表皮结构（400×）。
C. The epidermis (400×).

12. 角质层	12. Stratum corneum
13. 颗粒层	13. Granular layer
14. 棘细胞层	14. Stratum spinosum
15. 基底层	15. Stratum basale
16. 真皮乳头	16. Dermal papilla

图 11-1　皮肤
Fig.11-1　Skin

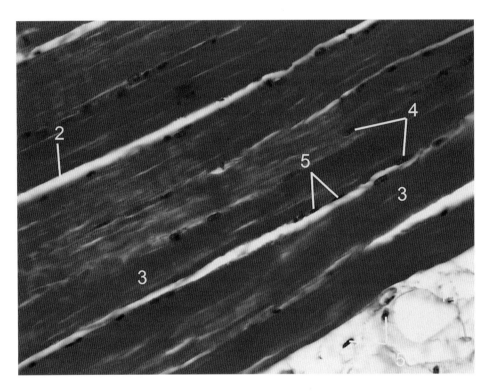

骨骼肌纵断面（400×）。
The longitudinal view of skeletal muscle (400×).

1. 肌束膜	1. Perimysium
2. 肌内膜	2. Endomysium
3. 骨骼肌细胞	3. Skeletal muscle cell
4. 骨骼肌细胞核	4. Skeletal muscle nuclei
5. 结缔组织细胞核	5. Connective tissue nuclei

图 11-2 骨骼肌

Fig.11-2 Skeletal muscle

图书在版编目（CIP）数据

骆驼组织学彩色图谱/苏布登格日勒，李海军著．—北京：中国农业出版社，2021.11
国家出版基金项目　骆驼精品图书出版工程
ISBN 978-7-109-28906-2

Ⅰ．①骆⋯　Ⅱ．①苏⋯②李⋯　Ⅲ．①骆驼-动物组织学　Ⅳ．①S824.1

中国版本图书馆CIP数据核字（2021）第221151号

中国农业出版社出版
地址：北京市朝阳区麦子店街18号楼
邮编：100125
丛书策划：周晓艳　王森鹤　郭永立
责任编辑：周晓艳　王丽萍
版式设计：杜　然　责任校对：吴丽婷　责任印制：王　宏
印刷：北京通州皇家印刷厂
版次：2021年11月第1版
印次：2021年11月北京第1次印刷
发行：新华书店北京发行所
开本：787mm×1092mm　1/16
印张：11.5
字数：290千字
定价：188.00元

版权所有·侵权必究
凡购买本社图书，如有印装质量问题，我社负责调换。
服务电话：010-59195115　010-59194918